WHO
REALLY
WON
THE
SPACE
RACE?

WHO REALLY WON THE SPACE RACE?

THOM BURNETT

COLLINS & BROWN

First published in 2005 by Collins & Brown
The Chrysalis Building
Bramley Road
London W10 6SP

An imprint of **Chrysalis** Books Group plc

Produced by Conspiracy Books
PO Box 51726, London NW1 9PT

British Library Cataloguing-in-Publication Data:
A catalogue record for this book is available from the
British Library.

ISBN 1-84340-290-4

The author and publishers have made every
reasonable effort to contact copyright holders. Any
errors that may have occurred are inadvertent and
anyone who has not been contacted is invited to
write to the publishers so that full acknowledgement
may be made in subsequent editions of this work.

1 3 5 7 9 8 6 4 2

Printed and bound in Great Britain by
Creative Print & Design (Wales), Ebbw Vale

Contents

Introduction

On October 4th, 1957, America's self-image, that of being the most technologically advanced nation on earth, was shattered by the successful launch of a Soviet satellite months ahead of its own satellite program. As *Sputnik* beep-beeped its way across the United States, and across the world, the public response to its orbit shocked the Eisenhower administration and delighted Soviet Premier Nikita Khrushchev.

Comedian Bob Hope told his TV audience: "Well, let's not get too upset. That simply means that the Russians' Germans are better than our German scientists." This joke was reinforced four days after the entry of *Sputnik* into space, history, and the Western language, when President Eisenhower gave a White House press conference. When asked why the Russians had beaten the United States into space, he attributed it all to the fact that in 1945, the Soviets had captured all of the German rocket scientists at Peenemünde.

As this book will show, that presidential statement was deceptive. Technically speaking, what Eisenhower had said was true: Peenemünde had fallen to the Russian Red Army, so all the rocket scientists in the area would have been captured. But his statement ignores the fact that the best of the scientists had deserted Peenemünde many months before. These Germans had then surrendered themselves to American forces and been transferred to the US after the war. Yet, with all that talent, heavily publicized by the leader Wernher von Braun's television appearances on the *Wonderful World of Disney* in 1955, America still lost the space race.

Conspiracy theories started to appear soon after the launch of *Sputnik* as the American public struggled to come to terms with the psychological blow of being beaten. Ex-President Truman blamed the virulent communist witch-hunt mounted by Senator Joseph McCarthy in the early 1950s for removing the best scientists. Future President John F. Kennedy started campaigning on the basis that there was a "missile gap" between Eisenhower's run-down ballistic forces and the Soviet's evidently superior ones. Although this was inaccurate, the popular perception of military inferiority would help launch Kennedy all the way into the White House after the next election. The *US News & World Report* claimed that Russian spies had stolen US rocket secrets. Others even claimed that *Sputnik* was a fake, part of a great deception by the Soviet Union.

Senate investigations into the reasons why the race had been lost soon revealed that a US Army missile designed by von Braun and his team of Germans could have launched an American satellite one year before *Sputnik*, but was denied the opportunity. This book will uncover the secrets behind this decision and expose the conspiracy at the heart of the US satellite program.

As America struggled to get into space, the Soviet Union launched yet another *Sputnik* on November 3rd, 1957, this time with a dog on board. The disastrous launch pad explosion of the US's Project VANGUARD then compounded Russian prowess and American impotence. The world's press referred to it not by its proper name, but by names of ridicule such as Flopnik, Stayputnik, and Kaputnik. It was only after this embarrassing failure that America finally allowed "its Germans" from Peenemünde to do the job they could have done back in 1956.

Using declassified intelligence documents from the United States and Britain, and the opening of the Russian archives, this book will show how little the German rocket scientists actually contributed to the winning Soviet team. It will raise doubts about the significance of Bob Hope's joke and, more importantly, Eisenhower's excuse.

PART I: NAZI VENGEANCE WEAPONS

Dreams

In his 1948 autobiography, *Crusade in Europe*, General Dwight D. Eisenhower admitted that if the German V-2 ballistic missiles had been ready six months before their actual deployment—the first V-2 was fired in September 1944—the Allied invasion of Europe, Operation OVERLORD, would probably have been cancelled. Thousands of ballistic and cruise missiles targeted at the assembled forces on the south coast of England would have proved a complete disaster for the Allies—and the course of world events could well have been changed. Why were the V-2s not launched earlier? In March 1944, six months before their deployment, and three months before the Allies fought on the beaches of Normandy, an extraordinary event took place at the very heart of the V-2 program. It was an event which would explain why Eisenhower's forces were spared the horror of Hitler's *Vergeltungswaffen*, his "Vengeance Weapons."

In the early hours of March 15th, 1944, General Walter Dornberger was awakened by the ringing of his bedside telephone. It was an urgent message from General Buhle, the Chief of the Army Staff, informing him that he was required to attend a conference at the Führer's headquarters in Berchtesgaden with the chief of the Oberkommando der Wehrmacht (OKW), Field Marshal Keitel, later that day. Setting off at eight o'clock in the morning, by his own account Dornberger braved "snowstorms, icy roads, and the havoc of a heavy air raid on Munich the night before" to reach Berchtesgaden by late afternoon. Booking into the Berchtesgadener Hof, Dornberger did not have to wait long for General Buhle to visit his room and tell him why he had been summoned to the Führer's headquarters.

At exactly the same time as he had left his quarters at Schwedt on the Oder, eight o'clock that morning, two of his most important rocket engineers at Peenemünde had been arrested by the Gestapo and charged with sabotage of the V-2 project. Their names were Wernher von Braun and Helmut Gröttrup, names which would become famous in the later missile programs of the United States and the Soviet Union, but which, in March 1944, were names at the very center of the Nazi missile program. Dornberger could not believe his ears: these men were the last people he would ever consider to be saboteurs. Their commitment to the research and development of the ballistic missile was total. The arrests by the Gestapo must have been a mistake, or, much worse, a deliberate conspiracy against the Peenemünde organization. General Buhle could tell him nothing more that evening. He would have to wait until he met Field Marshal Keitel the next morning.

Dornberger spent a sleepless night going over and over the various scenarios. He had known von Braun since the early 1930s and the days of the *Verein für Raumschiffahrt* (VfR), the Society for Space Travel. Under the leadership of the Transylvanian theoretician Hermann Oberth, this group of amateur rocket enthusiasts used to meet in the back room of a restaurant in Breslau, and rented a test firing facility in the Berlin suburb of Reinickendorf for a sum of ten Reichsmarks per annum. In the spring of 1932, Dornberger, then an artillery captain, had disguised himself in civilian clothes and visited the VfR to assess the abilities and achievements of the group. One particular young engineering graduate had impressed him—a "tall, fair young student with the broad, massive chin" named Wernher von Braun. When Dornberger was ordered to carry out rocket testing at the army's proving ground at Kummersdorf, he recruited von Braun as a technical assistant, along with several other VfR members. Such was the "astonishing theoretical knowledge" of the young man that Dornberger arranged for him to study for a PhD in Physics at the Friedrich-Wilhelm University.

The time spent at Kummersdorf had seen the development of liquid-propellant rockets of increasing size and power. Dornberger and his team of rocket amateurs were becoming not only professionals but also the foremost research and development team in the world. The Soviet equivalents of the amateur rocket societies were suspected of plotting the overthrow of Stalin through the use of missiles, and suffered terribly through the period of the Great Purge. The American interest in rocketry, pioneered by Dr Robert Goddard with the world's first liquid-fuelled rocket launch in 1926, failed to attract any serious investment from either the US armed services or the government, so the team at Kummersdorf soon dominated the new science. The field of military rockets was so new, in fact, that it was not covered by the restrictions applied to post-World War I German rearmament by the Treaty of Versailles, which made it an attractive proposition to the German Army—but, strangely, not to its leader, Adolf Hitler.

Hitler's relationship with the missile program had always been an unusual one. The Führer could get very excited about new tanks, but for some reason not about rockets. Whenever Dornberger had shown visiting dignitaries around Kummersdorf, they were all naturally awed by the power and explosive nature of the rocket engines. Not so the Führer: although he had a great interest in the technology of war, he was the only person that Dornberger knew who showed no interest in such machines. Dornberger could never fathom it out. When he showed Hitler around the test site in March 1939, he, Dornberger, did all the talking. It was almost as if the Führer was bored. Neither the roar of a blue jet of gas nor—despite thick wads of cotton wool—the painful vibration in his eardrums elicited a response. Dornberger had lain out a cut-away model of the proposed V-2 rocket on wooden tables, with the related components painted in the same color for ease of identification. Von Braun had explained how the entire weapon system would work and yet all Hitler would do was have a close look at the cut-away then turn away, shaking his head. He was a little more communicative over lunch in the mess, as if partially re-invigorated by his mixed vegetables and glass of Fachingen mineral water, but never-theless not optimistic about the program. It was then that Hitler mentioned Max Valier, a former member of the VfR.

Valier had been an engineer at the Heylandt Works at Bretz near Berlin and Hitler had got to know him fairly well at Munich during the early days. Valier had explained all about rockets to him then, promoting the advantages of liquid-propellant motors in racing cars. Unfortunately for Dornberger's case, in May 1930, Valier had disregarded safety precautions and blown himself up. Hitler called him a dreamer.

That word "dreamer" was applied to anyone at the forefront of technology. Dornberger had to tactfully inform Hitler of this, but it only led to more pessimism. He pointed out to the Führer that great names such as Valier, Oberth, and Goddard were to space travel what Lilienthal had been to the airplane, and Zeppelin had been to the airship—and both of these had required a long period of research and development. Before Dornberger could regret includ-ing the Transylvanian Oberth or the American Goddard in his list of greats, Hitler took him to task on the example of Count Zeppelin. In the Führer's expert opinion, the airship had not been a great invention. Dornberger asked him if he had ever been on board a Zeppelin airship.

"No, nor shall I ever get into an airship. The whole thing always seems to me like an inventor who claims to have discovered a cheap new kind of parquetry which looks marvellous, shines for ever and never wears out. But he adds that there is one disadvantage. It must not be walked on with nailed shoes and nothing hard must ever be dropped on it because, unfortunately, it's made of high explosive. No, I shall never get into an airship." With that, Hitler left. That had been in March 1939, before the move to Peenemünde. If only Hitler had come on board the rocket program then, events would have turned out very differently.

Yet, Hitler did make a curious statement later that year at a rally in Danzig which was picked up and fretted over by British intelli-gence. He had remarked that if the war continued for another four or five years, Germany would have access to a weapon now under development "that will not be available to other nations." British intelligence would have done well to read the November 1939 edition of the US magazine *Astronautics*, which speculated that Hitler's speech was a reference to military rockets. The article reasoned that the previously strong German rocket society, the VfR, with its membership of over a thousand enthusiasts, had been disbanded in 1934, but that despite the American belief that the German experiments had stopped, it was more likely that the society's activities had been absorbed into the secret preparations for war. Instead of considering this, British intelligence became obsessed with far more outlandish, fantastical ideas.

Despite the Führer's lack of enthusiasm and his decision not to allocate it priority funding, Dornberger and von Braun went ahead and supervised the transfer of the missile program from Kummersdorf to the newly constructed test site at Peenemünde on the Baltic coast. The nature of these secret weapons necessitated the tight security surrounding the new experimental station where all employees, including the seventeen thousand personnel working on the V-2 project, were vetted by the SS and the Gestapo. In addition, all the civilian contract scientists, engineers, and technicians came under the control of military law. For this reason alone, Dornberger found it incredible that von Braun and Gröttrup could have been accused of sabotage.

Helmut Gröttrup was the deputy director for rocket guidance systems at Peenemünde. Dornberger did not know him as well as he did von Braun, but his loyalty was beyond doubt. Due to its position and the secrecy surrounding its weapons programs, Peenemünde had developed into a closed community with many employees living with their families on the base. Gröttrup was a fine example, living with his wife Irmgard and his two young children. Irmgard was of a higher social class than her engineer husband and found she had much more in common with the young Wernher von Braun, who was the son of a wealthy Prussian banker and possessed the title of Baron. Through his wife, Gröttrup entered von Braun's elite circle of friends. However, General Dornberger could not see that group conspiring to sabotage the V-2 program.

The development of the V-2 had certainly been hindered by the low priority status given to it by Adolf Hitler. The bombers of the Luftwaffe drew the bulk of the defense spending, prompting Dornberger to argue that at a cost of forty thousand marks each, his V-2s were more economical than the estimated one and a quarter million marks it cost to build a bomber and train its crew. He was using official Luftwaffe statistics to back up his claim, since the bombers only had an average operational life of between five and six trips over England. Even when the Luftwaffe's defeat at the Battle of Britain in March 1941 proved his point, Dornberger found that Hitler was still unsure of the potential of the V-2.

The turning point in the V-2's fortunes came on October 3rd, 1942, at Peenemünde. From ninety feet up, on the flat roof of the launching site assembly building, photographers, and, more importantly, film cameramen, were ready as the loud-speaker announced "X minus 3. Counting off." The pictures were also being sent to television screens which showed, in Dornberger's own words, the "slender, perfectly proportioned body of the rocket, lacquered black and white, and glittering in the clear sunlight." The nose was shaped "like a rounded Gothic arch" and oxygen vapor formed "evanescent little balls of cloud" on contact with the moisture of the surrounding air.

It is interesting that Dornberger describes it in such a way, preferring to emphasize its romantic feminine beauty as opposed to the masculine, phallic nature of what was basically a weapon of war. That day's launching would cast aside all doubts that the rocket was a weapon of war. Up until that point, there had been two false starts, when the missile had hardly left the ground at all. There had been other cancellations too, usually at the last moment, due to technical faults. The frustration had built to such high levels that even Dornberger was getting pessimistic.

But October 3rd was to change all of that. It would go down in history as the start of the ballistic missile age and break the sound barrier at the same time. The film crews were in position to make a permanent record of this historic moment, and to provide proof to those people in power of what real power was about: enough thrust to lift a thirteen and a half ton rocket and send it nearly sixty miles into the upper atmosphere, smashing through the previous world height record held by the Paris Gun. A record such as this and the first supersonic missile were important to the Peenemünde scientists because they belonged to an international fraternity with a long history of rocket development, temporarily fragmented by the war, but still one which set limits and rejoiced when they were broken. Knowledge of achievements would, by necessity, be kept secret from other nations, but intelligence agencies would eventually piece together the evidence to show just how advanced the Germans were becoming. The cinematic record of the launch would have proved invaluable to the British in 1943, but its true value to Dornberger would not emerge for another year.

The last minute of countdown became known as the "Peenemünde minute." A smoke cartridge was fired into the sky and the direction of its green trail noted. Often wind was still a problem for launchings, but not on this occasion. Dornberger watched the television screen intently and noticed clouds emerging from the nozzle mouth, followed by sparks within the clouds, bouncing off the blast deflector and scattering over the concrete platform. The propulsion engineer had pulled the first of three levers to initiate ignition. The sparks quickly became a flame, then a stream of gas, and combustion lasted three seconds. Smoke darkened the screen. The second lever had been pulled for the preliminary phase. The cables and wood supporting the rocket were cast off on instruction from the third lever. A turbo-pump forced 33 gallons of oxygen and alcohol per second into the combustion chamber and after one second, the thrust level rose to 25 tons, easily lifting the 13.5-ton rocket.

As the V-2 left the ground, it also left the television screen, but the film crews outside were continuing to track it as it ascended out of the Peenemünde forest and into the sky. The large black-and-white checks painted on to the sides of the missile were there for a reason. As the rocket ascended, the pattern did not change, proof that the missile was not turning on its longitudinal axis. By the time the tip of the rocket started to change from the vertical to the 45-degree diagonal ideal for maximum distance, the sound waves of the ignition had just reached the Peenemünde staff, covering the 1,500-yard distance in nearly five seconds. In less than thirty seconds from launch, the missile was recorded as breaking through the sound barrier itself at over 650 miles per hour.

Signals that bounced back off the rocket from a high-frequency transmitter with a directional antenna showed that the missile continued to accelerate and remain stable in flight. Five seconds later, it had broken Mach 2, twice the speed of sound, and was six miles high. A trail of white smoke suddenly appeared and many on the ground thought it had exploded. A strange zigzagging of the smoke followed which confused people even more, but the signals continued to show that the missile was on course. The strange phenomenon was due to condensation of the combustion gases at different heights and different speeds over 2,000mph. It became known as "frozen lightning."

After nearly one minute of flight, the rocket engine switched off automatically and the speed of over 3,000mph was recorded as yet another record. Using strong binoculars, Dornberger was still able to see his beloved V-2, just "a tiny dot glittering dazzlingly white at the end of a small, dark streak." It was now moving through practically airless space, the result of nearly ten years of work. Out of sight of the film cameras and the most powerful binocular periscopes, but not the radio transmitter, the missile re-entered the atmosphere, braking down to about 2,000mph because of the rapidly increasing air friction.

The Peenemünde scientists were unsure of what would happen next. There was a great probability that the missile would start to melt and be torn to pieces by the intense heat and friction—but this was not to be the case. They were not to be denied a successful end to the experiment. Flight-Captain Dr Steinhoff calculated that the impact zone was 125 miles away, eastwards across the bay of

Swinemünde and about twenty miles north of the Pomeranian coast. He immediately set off in a Messerschmitt to reconnoiter the area. As was usual, a green-colored stain in the water would signal the point of impact, and this would be reported to a motor launch, which would measure the position accurately.

The results came back that the V-2 missile had been fired a distance of 120 miles, with only a lateral deflection from the target of two and a half miles. Dornberger, after reflecting on the day's historic achievements to his assembled team in Peenemünde, launched into a speech of a type that would turn out to be controversial:

> The following points may be deemed of decisive significance in the history of technology: we have invaded space with our rocket and for the first time—mark this well—have used space as a bridge between two points on the earth; we have proved rocket propulsion practicable for space travel. To land, sea, and air may now be added infinite space as a medium of future intercontinental traffic. This 3rd day of October, 1942, is the first of a new era in transportation, that of space travel…so long as the war lasts, our most urgent task can only be the rapid perfection of the rocket as a weapon. The development of possibilities we cannot yet envisage will be a peacetime task.

Science fiction had become science fact, appropriately with a rocket which had emblazoned on its side, together with the black-and-white checks, the logo of the 1929 science fiction masterpiece, Fritz Lang's *Frau im Mond* (Woman in the Moon).

When Dornberger publicized the success of the launch to the powers that be in Berlin, he was shocked to discover that still no-one was interested. The V-2 had acquired a competitor in the shape of the Luftwaffe's Model Fi-103, a jet-driven air torpedo, popularly designated the V-1. It was launched from a catapult along an inclined concrete ramp. Dornberger knew all about it, since he had originally passed on the research and development to the Luftwaffe in 1940, and because its testing station was also located at

Peenemünde. As if that wasn't bad enough, Speer reported that the real reason Hitler was unhappy with the rocket was because of a dream.

That was in March 1943, five months after the successful launch of the V-2. In order to determine which of the rival programs to back, a Long Range Bombardment Commission was set up, and a visit planned for May 26th. The two projectiles would be put head to head in front of a large audience, which included Speer, Admiral Dönitz, and Col-General Fromm. The V-1 was cheaper and carried the same amount of explosive, but flew at a low speed of 350mph. Its catapult launch ramps were easy targets for Allied aircraft and the missiles themselves could be intercepted. They could be detected both by radar and by the distinctive sound emitted by the air-breathing propulsion engine.

Dornberger's V-2 was ten times the cost but flew at a final impact speed of 2200mph and therefore could not be heard since it was travelling at over three times the speed of sound. It was also unstoppable. Despite the fact that two V-1s failed during the demonstration while the two V-2s blasted off to reach distances of 120 miles, both programs were pronounced to be at the same stage of development. The Commission recommended that both should be continued and given priority funding. That should have been the signal that Dornberger had been waiting for, but there was still the problem of getting Hitler's signature.

At last, on July 7th, 1943, Dornberger and von Braun were abruptly summoned to the Führer's Headquarters. Speer had told Dornberger to bring with him the film he had made of the successful record-breaking launch and any other promotional material. So, with Dr Steinhoff at the controls of a Heinkel 111, the three Peenemünders set off into the fog and uncertainty. As far as Dornberger knew, Hitler had never even seen a photograph of the V-2 in flight. He would show the film and let von Braun, ever the salesman, do the commentary. The film had been shown to other audiences many times, and no-one could deny the power and the thrill of watching the lift-off. Dornberger trusted that Speer planned to let it work on the Führer. Landing at Rastenburg, they were driven to the Army Guest House, Hunter's Height. From there, after a long delay, they were transferred to the Führer's Headquarters, a complex of huts and concrete shelters in a woodland clearing. The complex had its own cinema, and the team set up the film, models, and charts, and waited. Eventually, the door opened and someone announced: "The Führer!"

Hitler appeared wearing a voluminous black cape, flanked by Keitel, Jodl, Bühle, and Speer, all of whom would figure in the later arrest of von Braun and Gröttrup. They sat down in the front row of seats and the lights went dim. The screen showed the giant structure of the static test stand being wheeled out of the great assembly hall, over 90 feet high, and rolled over to the great blast tunnel. Scenes demonstrating the ease with which the rockets could be transported by road and rail followed test firings of the engines, all of this accompanied by von Braun's commentary. Then the actual preparations for the launch, with a hydraulic crane setting the rocket vertically on the launch pad. Extreme close-ups gave different angled views on the lift-off, with slow motion also being used to squeeze every last cinematic thrill out of the event. The rocket was tracked up into the sky as far as possible, then the film switched to animation of the trajectory, indicating the record speed, height, and distance achieved on that day nine months before. A fast resumé of key scenes ended the film and the message "We made it after all!" was left on the screen. Von Braun finished his commentary and there was silence in the cinema.

Hitler just sat there, slumped in his chair, staring gloomily ahead, silent, almost in a trance. Dornberger decided to break the embarrassing silence, and this seemed to shock the Führer back into the present. As the general explained the technicalities of mobile launching and the requirements for the industrial production of the missile, backed up by models, charts, and maps, Hitler was highly attentive, very different from his previous performance. When Dornberger ran out of things to say, Hitler grabbed his hand and shook it.

"I thank you," said the Führer. "Why was it I could not believe in the success of your work? If we had had these rockets in 1939, we should never have had this war." Then with true clairvoyance, he said: "Europe and the world will be too small from now on to contain a war. With such weapons humanity will be unable to endure it."

And with that, Hitler at last authorized top priority to the V-2 program.

Dornberger knew why the Führer had not believed in the success of his work previously, but sensibly chose not to remind him of it, for it involved a dream and a strange occult belief which Dornberger found impossible to accept. Hitler believed in the Vril, an intelligent energy force that surrounded and protected the planet, and his dream had shown V-2s igniting the upper atmosphere and destroying this protective force field.

Now Hitler had overcome that fear, but as Dornberger spent the night in his Berchtesgaden hotel contemplating the arrest of von Braun and Gröttrup, the string of inexplicable disasters that had plagued the V-2 program must have worried him. Since the beginning of 1944, all launch tests had ended in failure with either the rockets exploded on the pads, or in flight. With the V-2 already in production, the attempts to identify and correct the errors were hampered by pressure from the Armaments Ministry. One particular problem was the explosion of the rockets near the end of their trajectories, just ahead of impact. By March 1944, the rocket team at Peenemünde was no nearer understanding what was causing the problems and this must have been misinterpreted from above as sabotage. But why accuse the key personnel von Braun and Gröttrup?

The next morning, Field Marshal Keitel told him in no uncertain terms that the two engineers, and several others including Klaus Riedel and von Braun's younger brother, Magnus, were being charged with crimes so serious that they were likely to be executed. When Dornberger pushed for a clearer statement of the charges against his men, Keitel informed him that the accused had been overheard talking about certain subjects which he found incredible for men of their supposed standing. Without learning of the exact nature of their treacherous talk at that point, Dornberger claims to have vouched for them unconditionally, thereby risking his own position.

Then, Keitel revealed the charge. Von Braun, Gröttrup, and the others had been overheard saying that it had "never been their intention to make a weapon of war out of the rocket." They had worked under duress from Dornberger himself in order to achieve the parallel purpose so close to their hearts, Keitel alleged. And what was that purpose? What was such a treacherous thought that the Gestapo had arrested them and was in the process of incarcerating them at Stettin prison awaiting eventual execution?

"Their object all along has been space travel."

Space travel. A desire to break free from the bounds of gravity and journey to the stars. Such a desire obviously went against what the Reich demanded of its rocket scientists in the short term, though at some point during the proposed thousand-year Reich one might assume that it would become acceptable. Dornberger himself had joined in the frequent chat about space travel, for it was a common topic of conversation after the day's work was done. The V-2 missile being developed at Peenemünde was capable of reaching very high altitudes, and future improvements to the design and power of the thrust engines obviously brought the day closer when attempts could be made to launch something into space.

Dornberger admitted to Keitel that he had often openly speculated on the possibility of space flight and therefore if his men were guilty as charged then he too must be guilty and arrested. Keitel attempted to explain the charge more succinctly. "The sabotage lies in the fact that these men have given up their innermost thoughts to space travel and consequently have not applied their whole energy and ability to production of the V-2 as a weapon of war." Still hardly grounds for their execution. Dornberger failed to see how things could have gotten so unbelievably bad so fast, until Keitel admitted that it was now out of his hands. "Himmler has taken over himself."

Himmler. Reichsführer-SS Heinrich Himmler. A man who had recently attempted to wrestle control of Peenemünde and the missile program from the army and put it firmly within his SS. Himmler was also head of the Gestapo. Dornberger immediately suspected that the arrest of his men was another ploy by Himmler. Dornberger's strongest defender was the armaments minister, Albert Speer, but he was presently recovering from a serious illness and Himmler had obviously chosen this time to strike. Dornberger tried to convince Keitel that Peenemünde personnel were subject to military law and therefore outside the Gestapo's jurisdiction. His men should be handed over to him and he would ensure that they were placed in military detention, awaiting a proper investigation of the charges. More importantly, these men were needed to carry on their work, such was the importance of their positions within the missile program and such was the urgency of rectifying the current design faults. Keitel was unable to help and his reasons were embarrassingly political. As the last link between the Army's Officer Corp and Hitler, he was painfully aware of the precarious nature of his position. Himmler and his Security Service were watching him, applying pressure on him to leave at any and every opportunity. He could not be seen to be any less zealous than Himmler and his secret police in matters such as these. Dornberger would have to take such matters into his own hands.

He requested a meeting with Himmler himself, but when Keitel attempted to arrange this by phone with Himmler's adjutant, the request was denied. Dornberger was instead redirected to the SS Security Office in Berlin where his query would be dealt with by SS General Kaltenbrunner. Embarrassed by his admission of political impotence, Field Marshal Keitel dismissed Dornberger with a request that the details of their conversation be kept confidential. Dornberger drove back to Schwedt on the Oder "in a white heat of rage." But the run-around by the SS was not yet finished. When he arrived the next day at the SS Security Head Office in Prinz Albrecht Strasse, Berlin, with his Chief of Staff, Lieutenant-Colonel Georg Thom, he was told that his contact, SS General Kaltenbrunner, was not available. Instead, he was received by SS General Heinrich Müller.

"I ask for the immediate release of the gentlemen so surprisingly arrested by the Security Service," demanded Dornberger.

Instantly he found himself being educated in the various departments of state security. "I beg your pardon!" replied Müller. "In the first place, the gentlemen have not been arrested but are being held in safekeeping for questioning by the police commissioner at Stettin. Secondly, the Security Service has absolutely nothing to do with it. As a general on the active list you should surely know, by 1944, the difference between the Security Service and the Gestapo."

Von Braun and Gröttrup had indeed been taken into custody by the Gestapo, but Dornberger's confusion over the SS's role was understandable: Himmler had deliberately misled him by recommending that the case be dealt with through the SS Security Office in Berlin. Dornberger attempted to put forward the case for his men, but General Müller was non-committal. He would pass the details on to General Kaltenbrunner. However, Dornberger did manage to obtain permission to visit the prisoners at Stettin. Before he could leave, General Müller took a swipe at Dornberger's self-confidence: "You are a very interesting case, General. Do you know what a fat file of evidence we have against you here?"

By attempting to release his engineers from the claws of the Gestapo or SS or both, Dornberger found himself the subject of an ongoing SS investigation. Müller was indicating that they had a substantial amount of information relating to him that would require action, but not quite yet. Refusing to be fazed by the revelation, Dornberger challenged him to give him an example of the charges that were being levelled at him. Like his men, he seemed to be held responsible for delays in the development of the V-2. Dornberger was sorely tempted to lay the blame where it was due, at the feet of the Führer, but for obvious reasons he did not. However, as if reading his mind, Müller brought up Hitler's dream and Dornberger's reported comments on it.

"At the end of March last year you said at a meeting of your directors that the Führer had dreamed that the V-2 would never get to England. You said you were powerless against a Führer's dreams."

Müller went on to accuse Dornberger of exercising a pessimistic influence over the rest of the workforce. What zeal and enthusiasm existed among his senior staff members would have been severely dented by this expression of defeatism, he said. That in turn, using the same reasoning used against von Braun and Gröttrup, would, in effect, sabotage the rapid progress expected of the missile program at Peenemünde.

No consideration was given to the heavily defeatist sentiment of the Führer's own thoughts. Back in March 1943, the relatively low priority given to the missile program had caused the munitions minister, Albert Speer, to make repeated attempts to persuade Hitler of his oversight. Instead, what he received was an expression of second sight.

"I have dreamed," replied Hitler, "that the rocket will never be operational against England. I can rely on my inspirations. It is therefore pointless to give more support to the project."

This comment by Hitler had been reproduced in an official memorandum seen by Dornberger in Major-General Hartmann's office at the Ministry of Munitions. Dornberger told Müller that the memo had been typed in the specific large typeface used for Headquarters' communications involving the Führer. The reason for the large print was that Hitler was too vain to wear the reading glasses he was secretly dependent on. The existence of the specially constructed typewriters was a state secret and Dornberger's mention of the large print was an attempt to verify the truthfulness of his claims about the dream. If Hitler hadn't had the dream, then Dornberger's comments could rightly be viewed as sabotage. But Speer had confirmed it to him, he told Müller. The meeting with his directors at Peenemünde, which the SS were so apparently knowledgeable about, had not exactly happened the way Müller suggested. Dornberger had reminded his team that the rocket program had been full of obstacles right from the beginning and that, with determination, they had overcome every one of those obstacles. Hitler's dream was to be viewed as just the latest in a series of such obstacles. In fact, he portrayed it as the last obstacle in their path before their work at Peenemünde would receive the honor and acclaim that it deserved.

Dornberger mentioned to SS General Müller that the film he had shown to the Führer had successfully dealt with the problem of the dream. If he had been guilty of spreading defeatism amongst his team at Peenemünde, why would he have bothered to make the film in the first place? Müller did not reply, but the fat file was never pulled out by the SS, and action was not taken. Not taken, that is, until after the war when American and British investigators pulled out the file when SS documents became such hot items in the hunt for war criminals.

Wernher von Braun, on the other hand, was under investigation by the SS and he would spend the next two weeks in the prison at Stettin. To complicate matters for Dornberger, von Braun came under the jurisdiction of the SS, since he was in fact an SS major. Dornberger had even agreed to the young engineer's acceptance of the commission when it had first been offered. Not only that, but he had even forced the reluctant von Braun to wear the SS uniform once when Himmler had visited Peenemünde. Making the Technical Director an officer in the SS had obviously been one of Himmler's opening gambits in his attempt to take over the missile program. Von Braun was made fully aware of this in February 1944 when he was immediately summoned to Himmler's headquarters in Hochwald, East Prussia. The Reichsführer SS, looking to von Braun more like a country grammar school teacher than a monster baying for blood, told him: "I trust that your V-2 has ceased to be an engineer's toy and that the German people are eagerly waiting for it. I can well imagine what a pitiful position you are in: a poor inventor enmeshed by army bureaucracy! Why don't you come to us? You know that the Führer's door is open to me at any time, don't you? I shall be in a much better position to help you lick the remaining difficulties than that clumsy army machine!"

What Wernher von Braun said next has to be taken from the

perspective of hindsight, and bearing in mind that his is the only record of the conversation and, according to some historians, the only record of the meeting itself. Von Braun defended his commanding officer's position. General Dornberger was the "best chief" he could wish for and the delays were now technical ones, not red tape. The V-2 had to be compared with "a little flower that needs sunshine, fertile soil, and some gardener's tending...by pouring a big jet of liquid manure on that little flower, in order to have it grow, he might kill it."

According to von Braun's reminiscences, he was "politely dismissed." Telling the Reichsführer SS that his help in running Peenemünde would be like blasting a flower with liquid excrement was not the wisest thing to have said. If he indeed did say it, then it was no wonder that he eventually found himself in Stettin on conspiracy charges, no matter how trumped up they were.

As von Braun sat in his cell at Stettin, contemplating his future and with no indication of the charges against him, Helmut Gröttrup did likewise. Of lesser importance than von Braun, he was nevertheless a close friend and was Dr Steinhoff's deputy, and liaison to the Dornberger staff at Peenemünde.

According to the personal diary of General Alfred Jodl, Field Marshal Keitel's deputy in the Oberkommando der Wehrmacht (OKW), the event which led to their arrest was a private party at a hotel in Zinnowitz on March 5th, 1944, hosted by Hannes Lührsen, an architectural planner, and his wife. Lührsen was making use of his mother's hotel. Von Braun entertained his fellow guests—Gröttrup, Riedel, and his recently arrived younger brother, Magnus—with several classical pieces on the piano. Over drinks, the group chatted about space travel and it was this conversation that was being monitored by a local female physician who also happened to be a Gestapo agent.

Ten days later, at two o'clock in the morning, von Braun, Gröttrup, Riedel, Magnus von Braun, Lührsen, his wife, and even his mother, were rounded up and taken to Stettin prison. Eventually, von Braun was accused by a court of SS officers of making the statement about the V-2 not being intended as a weapon of war and space travel being the real reason for its development. Those were pretty lame charges, in von Braun's view, but what really worried him was the extra charge that he deliberately kept an airplane in readiness for a flight to England with all the plans of the V-2. He did have a private plane available but that was to service his wide-ranging duties within the missile program, not as an escape to the enemy. However, the charge would be a difficult one to disprove in the paranoid atmosphere of Berlin.

Jodl's personal records also show that there was another conspiracy charge for the group itself. They were accused of being a communist cell. It was claimed that Helmut Gröttrup and Klaus Riedel were members of either leftist or pro-Soviet organizations dating back to before 1933. Von Braun was linked to them through his close friendships, especially with Frau Gröttrup.

There was no denying that there was a Soviet influence on the Peenemünde team, mainly through the theoretical writings of the "Father of Cosmonautics," the Russian Eduardovich Tsiolkovsky. In 1895, he published *Dream of the Earth and Sky*, in which he discussed the possibility of an artificial Earth satellite, and three years later he created a mathematical formula to describe what would be required for a rocket, or "reaction machine" as he called it, to escape the Earth's gravitational field. This was further developed into what is still known today as the Tsiolkovsky Equation, the first theoretical proof of space flight, published in his 1903 *The Investigation of Space by Means of Reactive Devices*. Von Braun possessed a German copy of this book with almost every page covered in his own comments and notes. Among the Peenemünde archives were copies of drawings by the Soviet missile designer Tikhonravov. The fact that they were Soviet classified diagrams would indicate that the flow of intelligence was from the Soviet Union to Nazi Germany and not the other way round, but Riedel at least had been a "fellow traveller" of the Communist Party. The case against Gröttrup must not be swayed by his later employment by the Soviets, since in 1944 he was a loyal member of the Peenemünde team and his reasons for siding with the Russians after the war will be explained in detail later on.

As the engineers languished in Stettin, Armaments Minister Albert Speer was lying in a bed in Berchtesgaden, en route to an Italian mountain resort to recuperate from his illness. What had initially been a severe knee infection soon required hospitalization. It was followed by exhaustion, fever, and a life-threatening pulmonary embolism. While being briefly quartered in Berchtesgaden, Speer was seen by Hitler. He took the opportunity to complain about the arrest of his Peenemünde men and the conspiracy lies spread by the SS. Showing great concern for his friend's health, Hitler promised to deal with the problem.

And so it was that, armed with an order from Hitler, Dornberger was able to enter Stettin and remove von Braun, with an exemption from punishment based on his crucial role and indispensability to Speer. The others were conditionally released soon afterwards, but Gröttrup had to remain under some form of limited house arrest until the end of the war. Klaus Riedel, the one with the strongest communist leanings, died in a car crash.

Himmler's latest attempt to wrestle the Peenemünde operation from the army and put it into his expanding SS had failed. It seemed to have been plotted after the failed attempt at recruiting von Braun to the cause. As Speer recovered in the luxury of the Italian mountain resort and then returned to Germany, Hitler would grumble to him about the "trouble he had gone to" in sorting out this case. It is a good example of how political infighting within a nation's forces can easily destabilize it to the advantage of its enemies. The problem with the Allies was their apparent inability to pounce on this, since at the time they were still none the wiser about what this "secret weapon" was. Just as Dornberger found it difficult to publicize his V-2's successes and the potential use of the weapon in Berlin, so the British were unwilling to believe the snippets of information that were coming their way.

The one person to benefit from the whole SS charade was, ironically, Wernher von Braun. His brush with the SS would stand him in good stead after the war when he was able to avoid war crimes charges. Although an SS major, he had been imprisoned and inter-rogated by the SS and Gestapo for activities against the Reich. This would prove to be enough to brush aside any doubts about his true allegiance: it was not to the Reich but to the technological goal of space travel. After all, it was written down in black and white in the SS records, for investigators of other nations to see after the war. The Peenemünde team were dreamers, pushing aside the unfortu-nately necessary nightmares of war and weaponry, and maintain-ing their sanity by reaching for the stars.

At least, that was what they wanted the Americans to believe.

The Oslo Report and the
Failure of British Intelligence

Conspiracy theory fits well into the shadowy world of intelligence, where every piece of information received from foreign agents has to be screened for disinformation. A paranoid culture is developed in order to prevent a security breach caused by some clever act of deception on the part of the enemy. This can lead to very high levels of scepticism. Invariably the result is intelligence agencies dismissing real goldmines of information just because they appear too good to be true. A totally valid piece of intelligence might be thrown away because conspiracy theory rules the day. There are real conspiracies out there, especially during wartime, when shadowy groups are trying to sneak past your defenses and destroy you from within. The Allies were aware of this, because they themselves were carrying out similar covert operations against the Germans. One prime example of a goldmine ignored was the case of the Oslo Report.

On November 4th, 1939, Captain Hector Boyes, the British naval attaché in Oslo received, out of the blue, an anonymous letter offering him a secret report on the latest technical developments of the German military. To receive this report all he had to do was send a coded message over the radio, in the form of an altered preamble to the normal German language evening news broadcast on a certain date. The signal would be the introduction: "Hullo, hier ist London." The voluntary spy would take this as meaning that the British wanted to learn more of what he had to give. No money was ever mentioned. When the signal was duly broadcast and the information received, it claimed to reveal the development of two different kinds of radar, a large rocket, and a rocket-driven glider bomb. These last two were being tested at a secret establishment at Peenemünde.

With hindsight, the quality of the seven-page Oslo Report was exemplary. Yet the intelligence was not believed. Why? The principal reason was that the British Admiralty thought it was too good to be true and therefore had to be a devious plant by the Abwehr, the German intelligence service. The fantastical claims were written by psychological warfare experts to scare the British.

The one lone voice raised in support of its validity was that of Dr R.V. Jones, the newly appointed scientist in the Air Ministry's Intelligence Branch. It was he who had the unenviable task of opening the parcel in the first place. It contained a proximity fuse firing tube, which, luckily, wasn't misconstrued as being an explosive device sent over as a booby trap. Considering the whole package was supposed to be an elaborate hoax, the anti-aircraft fuse was a very valuable gift. The Oslo Report contained a note inside which read: "From a German well-wisher."

To this day, the identity of that well-wisher remains a mystery. Some have speculated that it was the head of the Abwehr himself, Admiral Wilhelm Canaris. If this had been the case, though, he probably would have been identified long ago, since there would be no reasonable excuse for keeping such an act of heroism to the Allied cause a classified secret.

The problem with the Oslo Report was that, after it, three years passed during which no further intelligence reports came anywhere near touching on the same subjects. In reality, the research and development of those rocket weapons mentioned in the report were beset by delays, both technical and bureaucratic. Nevertheless, once the radar and rocket projects were on course, Dr Jones found himself constantly referring to the Oslo Report to see what was coming next.

The precise mention of Peenemünde as the secret site of the rocket development and testing should have been harder to explain away. British intelligence should have checked the area out, but Jones explained that the state of RAF photoreconnaissance was not as advanced at the start of the war as some critics have claimed.

The limited number of available planes were used for more important targets, assessed at the time, and even then, Peenemünde was out of flying range for aircraft based in England. As the first Oslo Report confirmations came through in the form of German radar, aerial reconnaissance was concentrated around locating those radar installations.

Unfortunately, whereas a scientist within the Air Ministry's Intelligence Branch was open to the ideas of large military rockets, another scientist in a far more important position was not. Professor Frederick Alexander Lindemann, Viscount Cherwell, was the paymaster general and scientific advisor to the Prime Minister. For most of the period of uncertainty about the German missile, Lindemann obstinately refused to consider the large rocket a physical possibility. Even the Nobel laureate, Sir George Thomson, scientific advisor to the Cabinet, could not convince him otherwise. Basing his view on the simple premise that the amount of fuel required to blast a missile from the Continent to the southeast of England would be prohibitively large, he decided that the Germans would never waste time and effort trying. Thomson counter-argued that it was precisely the kind of challenge that the Germans would rise to and meet. Despite criticisms from other scientists, Lindemann's opinion held sway with the Prime Minister. If the British couldn't build one, there was no way that the Germans could. It was to prove a dangerously foolish position to hold, and Lindemann, to his eventual embarrassment and ruin, held on for as long as he could. It was so irrational that in other countries and in other times, conspiracy theory would question which side such a man was on.

Crazy viewpoints appeared to be rife when Dr Jones was given the early task of determining exactly what Hitler had meant in that Danzig speech when he warned the enemy about Germany being in possession of a weapon which no other nation would have. Given access to the secret MI6 files, he spent two months reading through a wide array of fantastical weapons classified as "secret projects." There were the dreaded bacterial and chemical warfare weapons which were all too believable, but there were also death rays and engine-stopping rays. MI6 funding had supported death ray experiments which ended up existing only in the imagination of a Dutch conman. Goats tethered to posts awaiting disintegration survived to live another day. The MI6 officer responsible for the experiment's funding did report that the machine could find an alternative use as a fruit preserver. Jones would eventually write that the whole secret weapon panic was the result of a mistranslation of the Danzig speech by the linguists at the Foreign Office. He claimed that by a "secret weapon" against which there was no defense, Hitler had actually meant the Luftwaffe. This hardly stands up when compared with the *Astronautics* magazine's interpretation.

Peenemünde was first photographed by the Royal Air Force on May 15th, 1942, when Flight Lieutenant D.W. Stevenson was returning from a recon mission over Kiel. Heading for the Swinemünde area, he flew over the island of Usedom, where he spotted some strange buildings near an airfield. Turning on his cameras, he recorded these before continuing with his mission to photograph the destroyers off Swinemünde. When he returned to England, Stevenson's photography was sent to the Central Interpretation Unit at the RAF station at Medmenham, where the photographs were interpreted and evaluated by personnel drawn from all the services and, since its entry into the war, the United States.

The intelligence process at the CIU was divided into three stages and these phases were meant to pick out anything of use. First Phase was an immediate reporting of anything vital to ongoing operations, such as the movement of aircraft, the location of ammunition dumps, and the damage assessment following a bombing attack. Second Phase was less immediate, usually to be produced within 24 hours of a mission. It was to be an overall perspective on the day's mission. Finally, the Third Phase was more of a specialized area of photoreconnaissance, dividing material up into geographical areas such as ports and airfields, industrial plant sites, and possible top-secret weapons establishments.

Flight Lieutenant Stevenson's photography caused puzzlement among the Phase Two interpreters. There were massive ring-like structures in the woods near the Peenemünde airfield, which they logged as "heavy construction work" but beyond that, they couldn't deliberate too long since there were far more urgent analyses to be made on the photos of destroyers off the Swinemünde coast. The sortie photographs were then passed on to the Third Phase where the specialists in secret weapons were supposed to investigate the rings.

Constance Babington Smith, the head of the Aircraft Section, has appeared in several histories as the woman responsible for spotting "something queer" in photos of Peenemünde which resulted in the identification of the Luftwaffe V-1. She was also the woman who looked at the strange ring-like embankments in the woods and recognized that something was amiss. "I wonder what on earth they are," she thought. "Somebody must know all about them I suppose." And with that she dismissed the whole thing from her mind. Her later success with the V-1 completely covered up the extreme negligence she showed over the V-2 launch sites. Just because there was no-one else within the Central Interpretation Unit (CIU) at RAF Medmenham who offered an explanation of the rings beyond explosive testing zones, they were ignored and the photos of the first sortie flight over Peenemünde were filed away in the print library for the rest of the year.

At the end of 1942, a small group of officers from the Military Intelligence Department of the War Office descended on RAF Medmenham with rumors that the Germans were developing long-range rockets to be fired at Britain from across the Channel. Intelligence reports mentioned that the heavy rockets were to be transported by rail so the CIU was tasked to search through the archive prints for evidence of possible railway and launching site developments. Agents on the ground around Peenemünde had reported that the whole area had been sealed off from the local population and secret weapons were supposed to be under development there. Even with a second look through the Stevenson sortie collection, the "huge elliptical embankments" went unidentified. Not even the introduction of fresh blood in the form of a team of army interpreters under Major Norman Falcon could correctly identify the strange shapes at Peenemünde as anything to do with rocket testing.

On June 13th, 1942, less than a month after Flight Lieutenant Stevenson's photoreconnaissance mission, the V-2 was ready for its first launching. It would show a dynamic display of cartwheeling more akin to fireworks before crashing into the Baltic Sea less than a mile away. Two months later, the second attempt managed to reach an altitude of approximately seven miles before exploding and raining debris seven miles off shore. On that particular occasion, Albert Speer had a wonderful view of the launch and its failure as his plane approached the airfield at Peenemünde West. One has to wonder what the odds of the missile taking him out were—wayward missiles were an ever-present danger on launchings. On Heinrich Himmler's first visit to Peenemünde, Dornberger had to drive him to the scene of a V-2 attack on the rival Luftwaffe section of Peenemünde. A wayward rocket had crashed into a hangar killing several workers and leaving an enormous crater fast filling with water. Dornberger would have found the whole incident most unfortunately timed, with Himmler already applying pressure to remove Peenemünde from army control and place it within the SS.

The British Chiefs of Staff proposed that a single high-ranking coordinator should handle the search for, and plan the defense against the German secret weapons. The man selected was Duncan Sandys, one-time head of the experimental rocket battery in Aberporth, Wales, and more recently first the financial secretary of the War Office and then the joint parliamentary secretary at the Ministry of Supply. He was sufficiently well connected to deal with the ever-cynical Lindemann, mainly through the fact that he had married Winston Churchill's daughter and thus he had the ear of his father-in-law, the Prime Minister.

Slowly but surely the evidence for Peenemünde being the secret weapon testing ground began to accumulate and yet still Lindemann wouldn't believe that a large rocket was possible. As reports of the V-1 started to appear, Lindemann was willing to accept this and saw it as confirmation that he had been right all along. The

V-1 was not a large rocket. When prisoner-of-war interrogations and eavesdropping revealed snippets of information about the V-2, Lindemann would dismiss these as being nothing more than mis-informed accounts of the pilot-less flying torpedo. In fact, the same was true of MI6 and other intelligence services. Despite the fact that there were intelligence reports concerning such developments, no-one seemed to consider the possibility that the Germans were devel-oping two missile systems at the same time.

Fritz Molden was an Austrian anti-Nazi deserter from the Wehrmacht whose father happened to be a professor at the University of Vienna. Constantly on the run, Molden had established a network of informants in Hungary, Rumania, Yugoslavia, and Czechoslovakia, as well as his native Austria. These informants were mainly friends of his father's, and several of them were scientists working on the secret weapons projects at Peenemünde. Committing every piece of information to memory and never car-rying incriminating notes on him, Molden would cross into neutral Switzerland by walking across the Alps. Rendezvousing with his Allied contact he would reproduce the information with phenom-enal recall, although the drawings of the two different missiles under development at Peenemünde were later described as little more than childlike. Nevertheless, these drawings found their way to the OSS and were passed on to British intelligence.

Molden was a very impressive agent in more ways than one. He managed to woo the daughter of his OSS controller and marry her after the war. His father-in-law was none other than Allen Dulles, the future director of the CIA. Dulles, operating out of Berne, was also receiving V-1 and V-2 intelligence from a contact close to Admiral Canaris at the Abwehr: Hans Bernd Gisevius. Dulles' public admission about this source in a newspaper interview in 1982 raises doubts that Canaris was the German well-wisher behind the Oslo Report, since surely Dulles would have known this if it were true. Some stories are just too great to keep hidden for so long.

In an attempt to get to grips with the British search for secret weapons, the newly appointed Sandys informed the Air Ministry to give it the highest priority. The army interpreters at the CIU were taken off Peenemünde and placed on the search for possible launch sites on the French coast. In a switch of services, an RAF team was tasked with keeping a close eye on Peenemünde. Finally, under the leadership of Flight Lieutenant André Kenny, the earthworks in Stevenson's photographs were correctly identified as rocket testing stands.

Kenny went from RAF Medmenham to London to inform Sandys at the end of April 1943, nearly a whole year after the original photoreconnaissance mission had first spotted the strange elliptical embankments. In May, activity at the CIU picked up as more sorties were flown over Peenemünde. Kenny spotted an image of a road vehicle carrying a huge cylindrical object, which the CIU interpreters estimated to be thirty-eight feet long and eight feet wide. Could this be the elusive rocket? POW interrogation intelligence also produced a result: a captured German general talked about rockets being fired in the Baltic area, from Greifswalder Island toward Bornholm Island, near Peenemünde.

On May 17th, 1943, Sandys proposed that the mystery of the conflicting reports about secret weapons could be solved if not one but two weapons systems were under development. Finally British Intelligence were on the right track. It had only taken them three and a half years to realize that a large rocket and a rocket-driven glider bomb were being developed at Peenemünde. This was precisely the information given to them so freely by their mysterious German well-wisher in the Oslo Report.

Another sortie was sent on June 12th, 1943, to find confirmation. The results were analyzed by CIU but nothing was found. Six days later, exactly the same reconnaissance photos were looked at by Dr Jones and he managed to spot something that the RAF team had inexplicably missed—a railway car transporting a V-2. The "whitish cylinder measured an estimated thirty-five feet long and at least five feet in diameter. It had a bluntish nose at one end and fins at the other." Jones had managed to find his target all by himself. A further sortie was flown four days later and this time, two rockets were shown clearly enough for even the RAF team to spot. This was also the series of photographs that earned Constance Babington Smith her title "Miss Peenemünde." It was the first time that the V-1 was captured on film.

But Lindemann's influence still permeated the interpreters at the CIU. Taking his lead, they claimed that the rockets displayed on the recent photos were not real, but dummies—fake missiles being transported about Peenemünde to deceive the RAF into believing that the wonder weapons were real. Peenemünde itself was probably an elaborate decoy, they ventured, to draw attention away from the real targets, wherever they may be. The British had great experience in the art of military deception, having become quite adept in the manufacture of dummy vehicles, or disguising tanks as trucks and vice versa. Dummy aircraft were stationed at dummy airfields to mislead the Luftwaffe into bombing the wrong targets. With such a high level of deception, even to the extent of using stage magicians as technical advisors, it was obvious that the Peenemünde rockets would conjure up similar thoughts.

Dr Jones tried to reason with Lindemann that if the rockets were dummies then the Germans were deliberately inviting the RAF to bomb Peenemünde. Now if Peenemünde really was a decoy itself, then it was an incredibly expensive decoy since intelligence from a completely unrelated source had come into his possession. A list of experimental stations had been compiled by a petty clerk in the German Air Ministry, giving instructions on the allocation of petrol coupons. Peenemünde was ranked above other stations which were undeniably important. Lindemann still refused to believe that a missile that large could be lifted off the ground using his state-of-the-art knowledge of British rocketry.

He was right in a way. The British could never have achieved it, mainly because they were ignorant of the type of fuel the Germans were using. Solid propellant development by the British rocket scientists Alwyn Crow and William Cook misled them into doubting the German technology. Having little or no experience of liquid fuel propellants was no excuse, however. The very fact that pre-war

German research had been in liquid, not solid propellants should have alerted them to the possibility. Even Goddard's American research had been along German lines, but still Lindemann and his solid fuel believers shook their heads. Crow and Cook were so wrong in their beliefs that they calculated the weight of the rocket to be up to a hundred tons with an eight-ton warhead. Such a monster missile would have been capable of killing four thousand people with a single impact in London, if only it could lift off the ground in Europe. Fortunately, Sandys was convinced that the rocket was real, and from then on it received the code-name BODYLINE.

Once Peenemünde was identified, the number of sorties was curtailed so the Germans would not suspect that the British knew what was going on at the secret installation. The last photos from authorized sorties had shown that the anti-aircraft defenses had been strengthened and smoke generators were in the process of being installed. Both were strong indications that the Germans were preparing for possible air attacks. They would not have to wait long.

Churchill was convinced from the photos that the threat from the secret weapons was immediate and needed to be dealt with before it became a serious problem. On the night of August 17th-18th, 1943, with a full moon in the sky to guide them, almost six hundred RAF bombers left bases in England and flew towards Peenemünde. Because of the distance, the bombers could not make use of British-based ground radar and the airborne H_2S radars were not always reliable for the type of precision bombing that this, Operation HYDRA, required. The full moon was therefore crucial to the mission.

The leader of Bomber Command, Air Chief Marshal Sir Arthur Harris, planned to improve the accuracy of the bombing by sending a small pathfinder force ahead of the main force. It was their role to locate the ground targets and identify them by releasing flares which would be easily visible to the main bomber groups. The master bomber would then confirm the targeting by flying low over Peenemünde. If the flares were accurate, he would radio the main force, which would commence the mass bombing.

There was plenty to go wrong with the plan. The Germans would be alerted to the pathfinder force's approach, they would have time to respond as the pathfinder marking with flares would be followed by the single master bomber fly-by inspection, before the arrival of the main force. The master bomber could be shot down before confirmation of the targeting, or the experimental radio-telephone could malfunction, or even be jammed by the German transmitters. Harris weighed up the odds and decided to go ahead with the plan, relying on the full moon and faith in the ability of the master bomber to stay alive long enough. Bomber Command knew that the nights were of sufficient lengths at that time of year to allow the moonlit conditions to aid such a long penetration of enemy airspace.

The success of the mission depended on the element of surprise. Harris did not want the anti-aircraft defenses to be ready and waiting for them or the Luftwaffe fighters in the sky. For this reason, security for Operation HYDRA was incredibly tight. The flight crews were not informed of the target until they were actually in the air. Eight Mosquito light bombers took part in a diversionary attack on Berlin, code-named Operation WHITEBAIT. It was hoped that the Germans would assume the large bomber force was following them, instead of veering off towards Peenemünde.

That night Wernher von Braun was entertaining a famous guest in the Hearth Room of the officers' club at Peenemünde. Her name was Hanna Reisch and she was the Luftwaffe's celebrated test pilot and the only civilian to hold the Iron Cross, First Class. The following morning she was due to make a test flight in the experimental Messerschmidt Me163 rocket-powered airplane. When the party ended, von Braun saw her to her car which then took her to her quarters at Peenemünde West. He decided to retire to bed in the bachelors' dormitory known as Haus 30. No sooner had he fallen asleep than the air-raid sirens sounded.

Hastily getting dressed, von Braun rushed over to the communications center and was informed that waves of bombers had been picked up by radar leaving England and approaching the area across Denmark and Schleswig-Holstein known as the "minimum flak exposure route" to Berlin. The bombers were believed to be on yet another of the frequent raids on Berlin, and therefore of no threat to Peenemünde itself. Once he heard over the radio that the Luftwaffe's new night fighters, Heinkel He-219s, had been dispatched to defend Berlin, von Braun decided to retire once again to bed. However, as he walked from the communications center back to Haus 30, he noticed that the defensive smoke machines had been activated. Then, suddenly, a flare lit up the artificial fog. Anti-aircraft guns opened up at Peenemünde West, Lake Kölpin, and Karlshagen to the south.

Von Braun managed to reach the concrete blockhouse nearest to Haus 30 just as the first bombs fell. Dornberger was still looking for his slippers, cursing the fact that his army boots were being cleaned. Walking carefully through broken glass and debris, he too made it unharmed to an air raid shelter, although an explosion ripped off the thick, steel door of the shelter. When an incendiary bomb hit von Braun's office, he risked his life to save as much documentation as possible. Research and development notes were of paramount importance to the secret weapons program and the need to keep them secure would figure greatly in later operations at the end of the war and during its immediate aftermath.

As the pathfinders approached the Baltic coast, three aiming points were to be hit by flares. The first of these, identified as F on the maps, was Peenemünde's housing estate where Air Ministry Intelligence claimed that the key personnel lived. It was not the original intention to bomb the living quarters, just to bomb the weapons establishment, but Duncan Sandys altered the targets at the last moment, at the objection of the Air Ministry. In his opinion the scientists behind the research and development were just as important as targets as the hardware. The second aiming point, designated E, was the area of Peenemünde's workshop and pre-production, while the third point, B, included the laboratories and offices.

This crucial targeting encountered problems right from the start as the pathfinders approached Peenemünde. They were to drop red markers along the north shore of Rügen Island but since the H_2S airborne radar failed to clearly identify the island and the presence of small clouds made visibility poor on the moonlit night, the red flares were dropped over the northern edge of Usedom Island instead. This produced a lethal error in the targeting of the housing estate. Further flares were wrongly dropped over a mile south of the intended target. When the master bomber, Group Captain John Searby, noticed the error during his fly-by inspection, he had to order a second set of flares, yellow ones this time, to be placed in the proper area. Unfortunately, once the final confirmation green flares were released by parachute over point F, one third of the bombers attacked the first set of red flares instead, dumping bombs and incendiaries on the Trassenheide slave labor camp instead of the housing estate. Instead of the cream of German scientists, Bomber Command managed to kill nearly five hundred Polish prisoners of war.

Two thirds of the first bombing run did succeed in aiming at the right target, but the delay had allowed those personnel to seek cover from the air raid, all except the chief of propulsion development, Walter Thiel, whose beach hut took a direct hit, killing him and his family.

Hanna Reisch could have test flown her rocket-powered Messerschmidt Me163 the next morning if she had wanted to since, for some strange reason, the Luftwaffe's experimental base at Peenemünde West had been completely overlooked. Since that was the development site for the V-1, and the CIU must have originally detected the doodlebugs on its runway, it is inexplicable how a massive bombing raid on Peenemünde could have ignored the obvious.

On Saturday August 21st, 1943, Peenemünde buried its dead. With ruthless wartime efficiency, 735 bodies were interred: single graves for the German civilians who were killed in the housing estate; mass graves for the soldiers and foreign workers. Placing the *Versuchskommando Nord* soldiers with the Polish labor camp workers was surprising. The Trassenheide camp had eighteen of its thirty huts destroyed in the raid, with many Poles dying in their sleep.

Those who tried to escape by climbing over camp fences tended to be blown apart there. Despite hitting fifty out of the eighty buildings making up the development works, Bomber Command was unsuccessful in destroying most of the important targets. The supersonic wind tunnel was untouched, as was the liquid oxygen plant. The unknown prime target, the Measurement Building, only suffered light bomb damage. Nearly six hundred bombers had been sent over to destroy Peenemünde and all they really achieved with their sixteen hundred tons of explosives and nearly three hundred tons of firebombs was a levelling of the housing estate and a massive reduction in the number of Polish POWs.

In a further twist of fate, the main effect of the Peenemünde raid was to force the Germans to disperse their missile testing program to other facilities, particularly a remote site in southern Poland, and straight into the arms of the SS. The location selected was Truppenübungsplatz Heidelager, the SS training camp near the tiny village of Blizna.

The V-2 would soon be ready for mass production and this would be handled in an underground factory being prepared at Niedersachswerfen, near Nordhausen in central Germany. A week after the Peenemünde raid, Himmler convinced Hitler that the tight security surrounding the employment of foreign workers in the actual production of the missiles should be relaxed. If the V-2s were to be produced in sufficient quantities to turn the tide once more in Germany's favor, concentration camp workers would have to be used. Himmler ordered SS Brigadeführer Major General Hans Kammler to take charge of the slave labor program, a controversial part of the V-2 project which went on to cause problems for von Braun and some of his fellow engineers in the years after the war.

In November 1943, the V-2 missiles were ready for testing at the new base in Blizna and the local Polish resistance forces, the Armia Krajowa (AK), were placing the former SS camp under increasing surveillance. German aircraft would be sent out regularly to locate the impact points of the missiles and radio the positions to mobile ground recovery crews. The speed with which the recoveries were carried out showed that the Germans wanted to keep their secret weapons secret. The AK set about trying to locate the likely impact zones with the intention of somehow beating the recovery crews to the prize, and a prize it would be if they were successful.

The opportunity eventually presented itself on May 20th, 1944, when an entire V-2 came crashing down into a riverside marsh. Polish soldiers of the local AK, the 22nd Infantry Regiment, rushed to the scene ahead of the German recovery crew and quickly rolled the relatively undamaged missile into the river and concealed its location through the clever use of a local farmer's herd of cattle. Driving the cows into the river stirred up the riverbed and muddied the water in time for a peaceful pastoral scene to greet the arriving search team.

The Germans searched the surrounding area and had to give up the quest. This left the Polish resistance with a heaven-sent opportunity to explore the workings of the wonder weapon. An enormous operation was set up to deal with the secret recovery and transportation of the rocket. A tractor had to be acquired to drag the missile out of the river and the steel cables for the task were only available in Warsaw. When local conditions were sufficiently secure, the AK dragged the missile out of the river and hid it in nearby woods. Extremely hard-to-find tools such as metal cutting torches were brought into the area and the V-2 was slowly disassembled. The parts and components were then taken to an old barn in Holoweczyce-Kolonia before eventually finding their way to Warsaw.

Colonel Utnik of the Sixth Bureau intelligence at the Polish Army Headquarters in London received the ongoing reports of the recovery operation and instantly doubted the veracity of the intelligence. What was it with wartime intelligence? Nothing was ever what it seemed. How could any news be viewed as real? Anyway, Utnik chose not to file the report away into obscurity, but to pass on the intelligence to the British. Major General Sir Colin Gubbins, head of the Special Operations Executive and a key member of the anti-secret weapons CROSSBOW Committee, urgently requested more information. A V-2 in Allied hands would greatly aid the war effort since countermeasures could be investigated. The need for intelligence on the missile received top priority and Colonel Utnik soon found himself swamped with requests.

In Warsaw, as scientists and engineers analyzed the missile components, information flow between Warsaw and London swelled to nearly three hundred coded messages per day. The Research Commission of the AK's Intelligence Bureau prepared a very professional and thorough four-thousand-word report on the investigation of the V-2, including diagrams, eighty photographs, and a sketch of the layout of the Blizna camp. The number of rockets fired and their recorded impact points were tabulated, showing the level of effort and patience the AK applied to the project. AK Intelligence also listed the factories used for the various components. The next stage was to ex-filtrate the V-2 components and parts, the Polish engineers who had studied them, and the four-thousand-word report, to England, where the British experts at the Royal Aircraft Establishment could finally learn what they were up against. Operation WILDHORN III was to be jointly planned by the Allies and the Polish underground movement headed by Tadeusz Bor Komorowski.

But it was not to be. The plan was refused by Supreme Headquarters on the grounds that the mission to gain scientific intelligence was not as important as the preparations for Operation OVERLORD—the invasion of Normandy. The AK forces were to be organized as part of the coordinated sabotage of German bridges and disruption of lines of communication. The V-2 stayed in occupied Poland for the time being. Instead, first prize in the race to provide the British with a V-2 was to come from an unexpected source.

Dornberger was at Blizna when the telephone call came from the Wolf's Lair, Hitler's headquarters, asking him if any missiles had been launched from Peenemünde recently. The Major General phoned Peenemünde and was told that neither a V-2 nor a V-1 had been launched from the sites. When he relayed this information on to the Wolf's Lair, he was told in no uncertain terms that someone must have been firing a missile from there since the Swedish government had made an official complaint to the Führer himself. Something resembling the description of a V-2 had broken up in an airburst over southern Sweden and fragments had landed near the town of Kalmar. The Swedes were in an uproar.

Dornberger got back in contact with Peenemünde and learned the truth. The missile should have been the world's first ground-controlled anti-aircraft rocket, codenamed WASSERFALL (Waterfall) but only the remote-controlled guidance system had been ready, so it was fitted into a spare V-2 rocket. The ground controller had lost visual sight of it when it disappeared into low clouds and panicked that it could be heading for a nearby coastal resort. To rectify the situation and make it safe, he had guided it northwards with his joystick over the Baltic Sea. Dr Steinhoff had gone off on his usual reconnaissance flight, but had not found any characteristic green marker dye in the water. The V-2 had been fully tanked up with fuel so it must have continued all the way across the Baltic to Sweden. So Dornberger then had to phone the Wolf's Lair again and explain that a V-2 had, after all, been fired at Sweden.

Would the Allies be able to discern its workings from the wreckage? he was asked. Dornberger had to admit that was likely, but in answer to the question about whether they would be able to learn how to jam the radio transmissions, Dornberger informed headquarters that the presence of a WASSERFALL guidance system in a normally unguided V-2 would certainly confuse the Allies. Nevertheless it was important to try and negotiate the return of the wreckage from the neutral Swedes. Dornberger was told to report to the Wolf's Lair immediately, where the Führer would vent his anger on him personally. By the time Dornberger arrived at Rastenburg, Hitler's temper had subsided and he was looking on the bright side of the crisis. It was not necessarily a bad thing that Sweden had learned Germany could launch rockets at its territory if it so desired.

Unfortunately for Hitler, and fortunately for the Allies, Sweden held firm and did not return the wreckage, despite several requests from the German Embassy in Stockholm. No-one had been killed in the impact, though a man had been thrown off his horse as the V-2 left a crater five feet deep and fifteen feet across. Souvenir hunters were quick to snatch up portable sections, leaving about two tons of wreckage for the Swedish authorities to transport back to Stockholm, where they carried out a study on the V-2. The technical branch of the British Air Ministry Intelligence sent two officers over to Sweden to examine the remains and negotiations were made to transfer them to the Royal Aircraft Establishment at Farnborough.

Once the deal was finalized, Lieutenant Colonel Keith Allen was allowed to fly a C-47 cargo plane belonging to the American Transport Command into Sweden, pick up the crates, and fly out again. Back in England the crates were taken to Farnborough and the contents reassembled as best as the British and American experts could. The first new discovery to be made was the smell of alcohol off the debris, which proved that the fuel used was alcohol. The dimensions of the partially reconstructed missile helped determine a more accurate estimate of its weight, thrust capability, and size of warhead. Lindemann's team estimates of a hundred tons were finally shown to be out by a factor of at least three, but even this was a massive error compared with its actual weight of just under thirteen tons. Based on the early liquid propellant work of the American pioneer Robert Goddard, equations were used to estimate the thrust and work out the likely range of the rocket. These too were overestimations, but they were far nearer than the previous ones.

Dornberger was right in anticipating the confusion which the WASSERFALL guidance system would cause amongst the wreckage investigators. It did not belong in the V-2, so the fear of a guided missile of that size was some consolation gained from losing a rocket to the Allies. The confusion did not last long however, when the Polish V-2 components and parts finally arrived.

As the Allied knowledge of the workings of the V-2 improved, so did that of the Germans. The problems were slowly being rectified and the main outstanding problem was airburst about two miles from impact. Dornberger suggested that von Braun should set up an observation post at the bulls-eye, which was not as suicidal an idea as it sounded. The inaccuracy of the V-2 was such that, over the distances of the test firings, no rocket had ever landed within two miles of the projected bulls-eye. By placing an observation post at the target, perhaps something could be visually spotted on the incoming trajectory which could explain why so many rockets exploded two miles out.

Von Braun agreed that dead center of the target zone would definitely be the safest place to be. The bulls-eye for the test was two hundred miles away from Blizna. As von Braun waited for the rocket to arrive, he wondered what problems in flight the rocket would reveal at a point in its trajectory never normally observed. His near brushes with death had previously occurred when rockets had exploded just after launch, not when they were coming down from the upper atmosphere at a speed of over 2,000mph. Von Braun saw a thin contrail appear in the sky and streak towards the target area. In fact it was to be the most accurate of the V-2s ever tested. Realizing that he was in danger of being hit by his own missile, von Braun dived for the ground but the impact blast threw him high into the air and into a nearby ditch. The one-ton warhead blasted out a crater only three hundred yards away from where he had been standing. The V-2 was ready for deployment.

It had taken the United States a long time to get into the war against the rockets, mainly because Churchill seemed to keep the intelligence to himself. Granted, Allen Dulles, and the OSS were providing snippets of information, but the raid on Peenemünde had been a British operation. However, things were to change rapidly once concrete ski-ramps were seen sprouting up all along the French and Belgian coast. As Operation OVERLORD was in the preparation phase, the ramps were associated with the V-1 rocket-propelled flying torpedo, and this put the American forces at risk in the south of England. In typical American fashion, Hollywood got involved.

The Joint Chiefs of Staff first discussed the problem posed by Hitler's secret weapons as late as December 1943. The intelligence that was available to the men in Washington created extreme fears about biological and chemical warheads being carried in the rockets. Three days before Christmas, General Marshall demanded an immediate report on the CROSSBOW countermeasures from the commanding general of European Theater of Operations, United States Army (ETOUSA), Lieutenant-General Jacob Devers. To the latter's embarrassment, he had to admit that he knew practically nothing. Marshall responded to the lack of intelligence by setting up a special committee to fill the gap.

On January 4th, 1944, Brigadier Napier of the British Ministry of Supply, who was in Washington on a fact finding mission of his own, was briefed on the state of American rocket developments by Dr Vannevar Bush. There was a certain irrationality about the low size of the V-2 warhead (then estimated to be about 1-2 tons) compared with the expensive cost of the rocket itself. Bush reasoned that the warhead of the long-range rocket must therefore be made from something other than high explosive. It had to be bacteriological. He then revealed to Napier that the Americans had been developing such weapons and had tested them successfully. Later the same day, Napier was summoned to meet Field Marshall Sir John Dill who complained bitterly about the fact that the Americans were pestering him for details on CROSSBOW, details about which he himself was largely ignorant.

The next day, Marshall's special committee met for the first time under the chairmanship of Major-General Stephen Henry. The intelligence so far accumulated indicated that the British had been painting a far rosier picture of the threat from the rockets. Exactly why the United States had been denied the truth was difficult to fathom. Thousands of miles away, in Marrakesh, at the top-secret talks on the planning of Operation OVERLORD, Lieutenant-General Devers was warning Churchill of the dangers of a German radioactive bomb which could be delivered by missile, producing radiation sickness and death over a two square mile area. Devers confided that the Americans had already experimented with such a device, so the Germans were bound to have considered it as well.

Churchill reported this back to Lindemann. "I do not know whether he is mixing up the possible after-effects on the lines of Anderson's affairs (I have forgotten the code-name)." "Anderson's affairs" was a reference to Sir John Anderson who was responsible for the British side of the atomic bomb project. The code-name which seemed to slip Churchill's memory was Tube Alloys. For a man who was so obsessed with the Enigma codes and the need for ciphers, to have forgotten the name of the biggest single project of the war may indicate the extreme stress he was under. While not suffering from radiation sickness, Churchill was reportedly unwell in Marrakesh.

Back in Washington, Brigadier Napier appeared before General Marshall's Committee. He was once again warned about the threat of bacteriological warfare, based on successful US trials in Canada, where Canadian goats and Canadian rabbits were sprayed with bacterial agents from low-flying US aircraft. The lethal effects lasted in the area for up to six weeks. General Marshall was very dissatisfied with the poor state of intelligence exchange on the CROSSBOW operation. It was his and the Committee's opinion that no serious attempt could be made to assist the British in the rocket countermeasures unless they were to receive all the information available.

The British Chiefs of Staff had a very different view of the threat from the special weapons. They preferred to deal with the high explosive warhead scenario and not get too obsessed with bacteriological threats. In order to calm the fears of their American cousins, the British Chiefs allowed the CROSSBOW intelligence to be shared.

Whether it was worth having is another matter. The Joint Intelligence Committee (JIC) estimated that the Luftwaffe might be able to launch by the end of March, a massive forty-five thousand pilot-less aircraft a month from the one hundred and fifty ski-ramps they had detected in France and Belgium. For once, Professor Lindemann's scepticism came to the rescue. In his view, this was way too high and the more realistic level of production was one and a half thousand per month. Even at this level of attack, Lindemann claimed that it would average out at one explosion within a mile's distance once a week for each Londoner. In another attempt to put the threat into perspective, Lindemann calculated that of every thousand launchings, only one would hit any specific square mile.

Statistics such as these were no comfort to the military planners of Operation OVERLORD. Alternative areas to Normandy would have to be considered for the invasion, or the ski-ramps would have to be continually bombed and new sites under construction detected quickly enough. Lindemann believed that there must be problems with the flying bombs since the ski-ramps had been constructed too prematurely. Hundreds had already been destroyed. In fact, Lindemann was correct in his thinking. The Volkswagen factory responsible for producing the V-1s was at a standstill; there were faults in the compass and the electrical fuse. The construction of ski-ramps continued in France, Belgium, and, surprisingly enough, the United States.

With admirable efficiency, General Marshall had ordered the reconstruction of the ramps at Eglin Field airbase down in Florida, together with the reinforced bunkers that appeared in the reconnaissance photographs. Thousands of civilian workers and army personnel set to work building exact replicas of the sites for bomber squadrons to practice and develop new bombing techniques. Hollywood film crews recorded the attack runs and the films were analyzed for the optimum results against the heavily fortified bunkers. The conclusion reached was that the US bombers were going to have to fly in low.

Armed with the cinematic evidence, just as Dornberger and von Braun had been when they went to convince Hitler, Marshall's staff officers flew to England and convinced General Eisenhower that the ski-sites could be effectively destroyed. The preparations for Operation OVERLORD continued unhampered by the rocket threat. Allied bombing of the targets now became top priority. In the seventeen weeks until D-Day, nearly twenty-three thousand tons of bombs were dropped on these ski-sites. The destruction was so easy that the Americans started to believe that the ramps were really an elaborate hoax, just as Lindemann had originally thought Peenemünde was. But detailed photoreconnaissance detected covert repairs being carried out at sites previously classified as abandoned.

As delays continued in the V-1 program, the Reich Propaganda Minister Dr Goebbels wondered if it would ever start. He had promoted various dates, such as the Führer's birthday, but that had come and gone. Hitler himself came nearest to guessing the date when he said, not as a result of a dream this time, that "The revenge bombardment is going to be synchronized with their invasion."

June 6th, 1944, D-Day, came and went too.

Photoreconnaissance continued to be carried out up and down the French and Belgian coasts. First Phase interpreters had spotted signs of modified ramps, and finally on June 12th activity was confirmed at six sites. At 4:18am the first V-1 exploded near Gravesend, twenty miles away from its intended target at Tower Bridge on the Thames in London. The second reached Cuckfield, the third Bethnal Green where it killed six people, and the fourth Sevenoaks in Kent. That was all. Two other missiles had been launched but they never made it across the English Channel. The first appearance of Hitler's "secret wonder weapons" was a huge anti-climax. Had the extensive bombing campaign destroyed most of the sites leaving a few that posed no serious threat? If this was the case, then there could be no further justification for the use of the three thousand Flying Fortress bomber sorties allocated for the complete removal of the ramps. The bombers could be redirected to the air operations over Normandy.

On June 16th, 244 missiles were launched at London. Eisenhower reacted by returning the status of the launch sites to top priority for Allied bombers "except for the urgent requirements of battle." At one point over forty percent of all the Allied bombing missions from England were directed at CROSSBOW targets. Some Allied air commanders believed that the rest of the air campaign was suffering from dangerous neglect, and on June 28th General Spaatz led the revolt. His strongly worded criticisms had no effect on Eisenhower who was determined to contain the damage inflicted on the British. On the same day, a V-1 exploded into the Air Ministry headquarters in the Strand, killing 198 people.

On July 11th, 1944, Dr Goebbels saw his first V-2 in action. Like Hitler before him, he was shown a film documentary, but this one was in color and had been commissioned by Albert Speer. The imagery overpowered the Propaganda Minister. "I believe that this missile will force England to her knees. If we could only show this film in every cinema in Germany, I wouldn't have to make another speech, or write another word. The most hardboiled pessimist could doubt in victory no longer."

Nine days later, the failed assassination attempt on Hitler in a Berlin bunker led to the arrest of von Fromm as one of the conspirators. His position as Commander-in-Chief of the Reserve Army had placed the secret weapons projects within his control, so his arrest opened up a vacancy in the power structure. It was one which was instantly filled by Dornberger and von Braun's nemesis— Reichsführer-SS Heinrich Himmler.

Scavengers

The V-2 wreckages salvaged from Sweden and Poland still left unanswered questions about Hitler's *wunderwaffe*, so it was with great interest that the Allies watched the advance of the Red Army through Poland to within fifty miles of the V-2 testing station at Blizna. It was only a question of time before the area was overrun. In a letter dated July 13th, 1944, Winston Churchill asked Josef Stalin for his cooperation in the location and retrieval of V-2 and V-1 parts and production materials which may be abandoned by the retreating Germans. In the spirit of Allied friendship, Stalin agreed to offer the assistance of his Red Army. In fact, he ordered the People's Commissar of Aviation Industry, Aleksey Shakhurin, to assemble a small task force of rocket experts which would enter Blizna as soon as possible with no intention of sharing any information with Churchill at all. Shakhurin recruited his team from the foremost rocket establishment in the Soviet Union, the NII-1. Among the members was the man responsible for the first ever liquid-fuelled rocket launch in the Soviet Union, Mikhail Tikhonravov. He too, like von Braun and Gröttrup, was a dreamer. He would become the man responsible for creating *Sputnik*.

Analysis of the Swedish wreckage led to a mistaken conclusion concerning the fuel used in the rocket. There were traces of hydrogen peroxide, but this was used to power the pumps, not the rocket itself. Hydrogen peroxide would have given the missile a far greater weight. If the rocket fuel really was that, the Allies reasoned, then industrial plants producing it would have to become priority targets too. For some strange reason, Peenemünde was identified as a hydrogen peroxide plant and targeted by the US Eighth Air Force. On July 18th, 377 bombers attacked its factories on that basis. On this occasion, in marked contrast to the earlier British raid of August 1943, the V-1 site at Peenemünde West was also hit.

The state of intelligence on the V-2 at that time was confused and best summed up by a comment made at the CROSSBOW Committee meeting held on the same day as the American bombing of Peenemünde. According to one of his sources, Major-General Gubbins, the head of the Special Operations Executive, had learned that the V-2 rocket was actually guided by a small man travelling in the nose cone who parachuted out to safety before impact.

As the Allied forces started to penetrate further into Europe, an important development in the acquisition of scientific intelligence took place. The Supreme Headquarters of the Allied Expeditionary Forces (SHAEF) created a fusion of soldier and scientist by setting up special units known as T-Forces. They operated at Army Group level and their role was to seize and secure any scientific asset that would be of use to the Allies. For example, the Twelfth Army T-Force was composed of personnel with language and scientific documentation skills, backed up by an armored infantry battalion, a tank destroyer company, combat engineers, and ordnance bomb disposal squads to clear the way. Identified by the large red "T" on their helmets, the forces would operate independently of the ordinary combat troops. They would wait until an area had been declared safe, then would systematically scavenge for anything of scientific and technical worth.

On July 27th, aerial reconnaissance photographs taken of Blizna showed that the weapons testing site had been abandoned in advance of the Red Army. Once again Churchill contacted Stalin and arranged for an Anglo-American team of armaments experts to fly from London to Moscow to accompany the Red Army into the area. Colonel Terence Sanders headed the mission with the American Lieutenant-Colonel John O'Mara as his deputy and a group of Russian, Polish, and German linguists backing up the rest of the team.

By the end of the month they were stranded in Teheran, wait-ing for the Russians to organize visas for Moscow. Delaying tactics would be employed to an excruciating extent all the way through the mission. Soviet cooperation and assistance in the investigation of a very serious threat to the Allied effort on the western front was nothing but a ruse. Stalin's constant desire to deceive the American and British intelligence services would lead to blatant acts of lying. Colonel Sanders was furious at the delay and cabled Duncan Sandys for diplomatic pressure to be applied.

Meanwhile, another 221 bombers attacked Peenemünde again as the imaginary threat of renewed hydrogen peroxide production led to some dangerous new developments in the art of bombing. The US Eighth Air Force, in an attempt to improvise its own flying bomb, stripped out old B-17s and B-24 bombers (affectionately termed Weary Willies) and packed them with high explosives. They were to be flown by flight crews who would then, like Gubbin's V-2 jockeys, bail out before the planes continued on their course by remote control. Known as Project Aphrodite there was nothing beautiful about the execution. Those planes that actually managed to get as far as Peenemünde missed their targets. One Weary Willie in particular never even made it out of England, its mid-air explo-sion destroying fifty-nine houses in the small town of Newdelight Wood and killing its two American crewmen. A photoreconnais-sance plane following it was almost brought down in the explosion, threatening the life of its pilot, the president's son Colonel Elliott Roosevelt. The pilot of the B-17 flying bomb was Joseph Kennedy Jr., the brother of the future President, and the details of his death were kept top secret until JFK gained access to the documents during the term of his presidency. In a cruel twist of fate, Kennedy was due to meet the family of his brother's crewman in Dallas on November 22nd, 1963.

Human intelligence revealed that the V-2 rockets were launched from a vertical position, so the search for exotic structures was aban-doned and on June 12th, 1943, all the archives of Peenemünde recon-naissance photographs were re-examined, including Stevenson's original photograph. Sure enough, once the interpreters knew what they were looking for, they found them all over the place. What had previously been interpreted as "thick vertical columns, about forty feet high and four feet thick" were re-measured and identified as vertical standing rockets. Ten sorties had produced evidence of no fewer than seventeen of the V-2 rockets standing upright in Peenemünde. The rocket transporters, the *Meillerwagen* trailers, were also now visible.

The whole episode, lasting just over a year, shows the difficulty of photoreconnaissance interpretation when strange objects have to be spotted and then their function guessed at. It is easier to spot that which is already known. Overhead reconnaissance would obviously have trouble catching a vertically positioned rocket. Rockets sighted earlier had been arranged horizontally, normally on a transporter. Finally confirming the method of launch only raised fresh problems and worries for the Allies. The operational V-2 launch sites would be very difficult to find, unlike the ski-ramp structures of the V-1.

Back in the Soviet Union, Stalin's team of rocket experts, led by Major General Fedorov, were not waiting in Moscow for the Colonel Sanders Mission to join them. Another team of experts would fulfill that role, eventually. Under "protection" from the NKVD (the Soviet secret police), Mikhail Tikhonravov and the small group of engineers from the NII-1 entered Blizna with the first Soviet troops on August 6th 1944, while Sanders was still stuck in Teheran.

Most of the laboratories and workshops had been cleared of equipment by the evacuating Germans, but Tikhonravov was still able to find a V-2 combustion chamber and parts of the propellant tanks. Everything of value was rounded up and loaded onto a Li-2 transport aircraft, then flown to Moscow under very tight NKVD security. A special group from NII-1 were then allowed to study the V-2 and this section became known by the top-secret designation RAKETA which, hardly cryptically, meant "Missile" in Russian.

Tikhonravov was joined by fellow Blizna veteran, Pobedonostsev, and, among others, a young specialist in control systems, Vasiliy Mishin. The latter had not been allowed to travel to Blizna since the

NKVD had a file on his father. Such were the workings of the Soviet state. Mishin required written permission to travel anywhere simply because his father had once heard a crude joke about Josef Stalin and had committed the crime of not reporting it. His father had served several years in prison for that. Yet the young Mishin would become an expert at obtaining information from the scraps of the V-2, and the NKVD relaxed its tight restrictions to include him in RAKETA.

Meanwhile, Colonel Sanders and his Anglo-American team were finally granted visas to enter Moscow. The next problem was that the Russians had forgotten to lay on a plane for the mission. Still stranded in Teheran, Sanders cabled London and Anthony Eden himself cabled the British Embassy in Moscow. With the full indignant force of the Foreign Office behind him, the Ambassador expressed his sincere hope that Marshal Stalin would still show the value of his personal interest in the matter of the missile situation.

The next day, August 8th, Sanders and his team were on a plane to Moscow, where yet more delays awaited them. According to the Soviet General Staff, the Anglo-American team could not be allowed to Blizna for the simple reason that it was still in the hands of the Germans. But the photoreconnaissance had shown that the Germans had deserted the area two weeks before. The area was still held by the Germans, came the reply. For a further ten days, Sanders and his men were entertained in Moscow until news of Blizna's occupation leaked out. The British Ambassador complained bitterly to Molotov about the need for Marshal Stalin to show that his promise to Churchill would be fulfilled. It would, came the reply.

While the RAKETA group at NII-1 continued to investigate the V-2 missile parts, another group was set up to consider the military uses to which the Soviet Union could put these wonder weapons. It was headed by Major General Lev Gaydukov, chief of the Communist Party Central Committee's department for artillery affairs. The V-1 cruise missile was deemed to be a waste of time and effort, so no Soviet replica would be attempted. The V-2, on the other hand, was unlike anything they had ever dreamed of.

On the day after the last V-1 was launched from France towards England, the Russians finally permitted Sanders to fly his team out of Moscow and into occupied Poland, though serious doubts were voiced about the very existence of the long-range rocket. The Sanders Mission eventually reached Blizna on the evening of September 3rd, 1944. Armed with the RAF reconnaissance photographs of Blizna and the surrounding area, the team spent the next few days locating the V-2 missile parts which both the Germans and the Russians had missed. Ten impact craters within five miles of the Blizna station had been the sites of misfires and Sanders was able to collect nearly a ton and a half of V-2 material, including a complete steel burner unit. The most important find was a forward fuel tank with an estimated capacity of 175 cubit feet, which would hold 3,900 kilograms of alcohol. The long stay with the Russians had educated the Anglo-American team about Soviet personal hygiene habits: one of Sanders' men searched the latrines and found important scraps of documents had been used as toilet paper. Once cleaned, these revealed a rocket test sheet, which identified two liquids: one was liquid oxygen and the other was called B-Stoff, of which the rocket's load was 3,900 kilograms. This tied in exactly with the fuel tank capacity if the fuel was alcohol.

The Sanders Mission was then flown to the V-2 impact sites at Sawin and Sarnaki, 100 and 155 miles respectively from Blizna. The V-2 wreckages had obviously been cleared up in a more efficient manner than the accidents back at Blizna. However, there were fragments to be found at the River Bug, the site of the AK retrieval operation during the early summer. The river, which had concealed the crashed V-2 so effectively from the German recovery teams, was at a lower level now and revealed some of its treasure.

In a final act of deceit, Stalin planned to get his hands on even that treasure. The ton and a half of V-2 fragments were crated up at Blizna and flown back to Moscow, where the Sanders Mission finally arrived on September 22nd. It had been a long, arduous operation, with the Anglo-American-Soviet team showing none of the fellowship and cooperation which would have been characteristic of allies, but displaying all that the future post-war world had in store. The Sanders Mission had been sent to Russia to gain advance scientific intelligence on the V-2 missile, but its investigations had been overtaken by events.

At 6:43pm on September 8th, 1944, the first V-2 landed on British soil, at Chiswick, killing three people and injuring seventeen more. Britain did not need to search foreign countries for V-2 wreckage any more—Hitler was delivering it. The Sanders Mission would have been at least partly consoled if they had managed to retrieve a vital part of the V-2 for analysis, but the crates forwarded on to London from Blizna via Moscow did not contain the V-2 fragments—they had been substituted with ordinary aircraft engine parts.

The ton and a half of Anglo-American property found its way to the RAKETA group's assembly room, where pieces of the V-2 were slowly being pieced together by Tikhonravov and Mishin.

Colonel Sanders' mission had failed in Poland because it had involved the hopelessly uncooperative Soviets. In the liberated areas of Western Europe, the T-Forces were left to their own devices. Within a short time they were making such great demands on SHAEF to assist in the extraction of scientific assets, that it complained to Washington and London. As a result, in late August 1944, a centralized agency was set up to organize the systematic exploitation of the Nazi scientific wealth. It was called the Combined Intelligence Objectives Subcommittee (CIOS) and drew support from the US Departments of War, Navy, Army Air Force, and State, together with the Foreign Economic Administration, the OSS, and Office of Scientific Research and Development. From the British side the participating agencies were the Foreign Office, the Admiralty, the Air Ministry, and the Ministries of Supply, Aircraft Production, Economic Warfare, and Fuel and Power.

With so many taking an active interest in the exploitation of German scientific assets, the operation was bound to become unwieldy and the CIOS acronym was soon being referred to as

CHAOS. A "black list" of targets was drawn up which CIOS determined were of prime military importance and teams of relevant experts were provided to SHAEF to investigate the areas once they were cleared of German resistance. From its London headquarters, CIOS sent its first team to Paris at the end of August. Within days, it had received a new target, based on human intelligence: the Central Works at Nordhausen—the place where the V-2s were being mass-produced.

As the Allies spread out from Normandy and overran the V-1 launch sites, the British became more confident that the retreating Germans were reaching the limit of their flying bombs' range. The estimated two-hundred-mile range of the V-2 was also being reached in everywhere except Holland. On September 6th, the British Chiefs of Staff concluded that "there should shortly be no further danger," and that all air attacks on suspected CROSSBOW targets could now be stopped. Such was the confidence that on September 7th Duncan Sandys held a press conference in London and uttered the unfortunate words: "Except possibly for a few last shots, the Battle of London is over."

The next evening, the Chiswick bombing signalled the dawning of the ballistic missile age. The sound of the explosion was heard before the sound of the missile approaching and twenty houses were destroyed. There was no press conference either announcing the arrival of the second type of Nazi wonder weapon or explaining the error of the previous day's statement. The news blackout was held in place for the next two months, during which two hundred of the "gas explosions" took place over southern England.

There was no defense against the rocket. The only hope was that the Allied forces would eventually push the German missile launchers in Holland beyond the two-hundred-mile range limit. Photoreconnaissance quickly discovered that the V-2s were being launched from inside the Haagsche Bosch Park in the Hague. The missiles were being transported by the *Meillerwagen* from the Central Works in Nordhausen to the dense woods of the Haagsche Bosch, then erected at different locations and fired towards England with a stunning success rate of over ninety percent. The *Meillerwagen* would then return to Nordhausen to pick up other missiles.

The power of the V-2 impressed the hell out of Colonel Gervais Trichel, Chief of the newly created Rocket Branch of the Ordnance Department of the US Army. He set up a firing range at White Sands Proving Ground in New Mexico in preparation for the expected acquisition and testing of these Nazi wonder weapons. The Ordnance Department contracted with the General Electric Company for a top-secret guided missile development project under the code-name HERMES. One of its key personnel was Dr Richard Porter. When the conditions were right for the exploitation of the V-2 factory at Nordhausen, Porter would be sent over to Europe with a team of General Electric engineers to join CIOS. For the actual acquisition of V-2s, the Ordnance Department would rely on some of their own scavengers.

After two months of detailed study and reconstruction of the missile fragments, Tikhonravov and his team of RAKETA engineers produced a lengthy report and recommended that the Commissar of the Aviation Industry, Aleksey Shakhurin, the man Stalin had ordered to investigate Blizna, should make every effort to rebuild and recreate the V-2 for testing, and develop a Soviet version for the military. Shakhurin was not impressed with the only valid aviation missile, the V-1, and he was not keen on putting his aviation industry's resources into a product which was really the concern of the Commissariat of Ammunitions. They had, after all, funded the development of the small solid fuelled Katyusha rockets. There was also a general belief that large missiles did not have an immediate future as major weapons of war. The Aviation Industry had full confidence in the future of bombers and fighters. With that decision, RAKETA was wound up and Tikhonravov and his team were sent back to their former duties at the NII-1.

Launch sites in Germany were too far to reach England, but they could be used to reach Antwerp, and after the start of the V-2 attacks in December 1944, von Braun and Dornberger were summoned to Schloss Burg, an old castle several miles outside Remscheid in the Ruhr Valley. Albert Speer had invited them to an extravagant award ceremony. Surrounding the castle in the dark forest were a number of V-2s aimed at Antwerp and each time one was launched, Speer decorated one of the recipients. The exhaust flashes and ground reverberations added atmosphere to the Wagnerian event. Von Braun was awarded the *Ritterkreuz zum Kriegsverdienstkreuz mit Schwertern* (The Knight's Cross of the War Service Cross with Swords). That was one launch site that the Allies never picked up on photoreconnaissance.

In the New Year, the original launch site at Peenemünde came within striking distance of the Russian Army. SS General Hans Kammler made the decision to evacuate the four thousand personnel that still worked there. They were to disperse to various locations, with the main group moving to the Herz Mountains. Dornberger relocated his headquarters to Bad Sachsa, Dr Kurt Debus moved his test stands team to Cuxhaven on the North Sea and Von Braun supervised the transfer of documents and equipment to Bleicherode, twelve miles from the Central Works at Nordhausen.

News came out of Blizna that more rocket fragments had been discovered, so after a few months of inactivity, the Soviet rocket scientists at NII-1 were again called on to fly into Poland. The mission was led once again by Major General Fedorov, but this time Tikhonravov was not involved. Mishin travelled to the Moscow airport intending to join this expedition, but as usual his father's crime, not reporting a joke about Stalin, prevented him from being allowed to travel. For once it was a fortunate restriction on his freedom since the aircraft transporting the scientific team to Poland was lost over Kiev, the crash killing everybody on board. Thus two key players in the future Soviet space program managed to escape death. Two further players had survived the death camps not of Adolph Hitler but

Joseph Stalin, and the paranoid beliefs of their leader kept them from the V-2 investigations until after the war.

Von Braun had escaped death on several occasions, for rocket research was a dangerous science. Dornberger had survived a gunpowder flash to the face. The future of the space race might have looked very different if any of them had died. As the Americans entered the European theater of operations, key personnel for the future of the US space program also survived near misses. John Medaris, the man responsible for the launch of America's first satellite, was an Ordnance Officer of the First Army in 1944 and was involved in the planning of the D-Day invasion. Once his headquarters was established in France, it was nearly at the receiving end of a stray V-2. Rattling the windows, the missile's blast had been absorbed by a nearby ravine and the headquarters' cook tent.

The man who grabbed the most missile war booty for the United States, and thereby launched the Americans on a fast track to the stars, also had a close encounter with the V-2's destructive power. Robert Staver was a twenty-eight-year-old major in the United States Army Ordnance, who was sent over to London in February 1945 by Colonel Trichel, Chief of the Rocket Branch. He was ordered to acquire some V-2 rockets and all the documentation he could lay his hands on, and to interrogate the German designers if possible. As he reported for duty at the operational headquarters in Grosvenor Square, he and his commanding officer, Colonel Horace Quinn, were thrown to the floor as a V-2 exploded overhead. With hot metal raining down on the square outside the window, Staver realized how close he had come to being killed by the very weapon he had been sent to search for.

The longer the major stayed in England, the more von Braun's rockets seemed to seek him out. While sleeping at his hotel near Marble Arch, he was blown out of his bed by a V-2 exploding at nearby Speakers' Corner. Once again, while on official Army Ordnance business to the Royal Aircraft Establishment in Farnborough, the site of many vital V-2 investigations in the days before the missiles started presenting themselves in England on a daily basis, Staver witnessed a direct hit on a Rolls-Royce Merlin engine factory warehouse.

During the two months he spent in England surviving near misses, Major Staver researched all he could on the V-2 as the Chief of the Rocket Section of CIOS. Although he had graduated from Stanford with a mechanical engineering degree and then spent three years with Army Ordnance's Research and Development Service, acting as liaison officer between its rocket development branch and the Jet Propulsion Laboratory at Cal Tech, Staver was still amazed at the technological advances made by the Germans over their American rivals.

The desire to obtain V-2 war booty was not one driven by greed but by desperation. The V-2s had to be seized and their designers found and made to reveal how they had been built. The state of American rocketry lagged pathetically far behind that of the men from Peenemünde. Staver's collation of intelligence under the auspices of the Combined Intelligence Objectives Subcommittee was, even at that early stage, purely for the benefit of the United States. The bulk of the intelligence came from British sources, but Staver was in London under strict orders from Colonel Trichel to compose a very important shopping list—one which would involve very little monetary purchasing. It was Staver's job to steal as much of the technological knowledge as possible and ship it back to the United States without the British noticing that it was gone.

In February 1945, the Big Three—Roosevelt, Churchill, and Stalin—met at the Yalta Conference, to divide up Germany. As the various zones of occupation were agreed, it became obvious that certain areas would be held by the wrong power when the antici-pated dates of victory and its attendant surrender of Germany occurred. As adjustments needed to be made, with one power retreating from land obtained by hard-fought battles and giving over to the correct occupying power, there lay opportunities to ransack the technological booty within the time agreed for transfer. These opportunities for plunder fell mainly on the American forces which had penetrated well beyond the boundaries established at Yalta.

Staver compiled two lists of targets: the Black List for major ones, and the Grey List for lesser ones. The lists were drawn up from British intelligence files on known scientists working for the various weapons centers. Large numbers of suppliers of parts were catalogued and aerial reconnaissance photographs of the production center at Nordhausen were provided in the belief that Staver was working for the benefit of the Allies, not just his own country. It would end up being yet another example of the gullibility of British Intelligence. There was no doubting that the British were very good at gathering intelligence on the V-2 once the Allied forces were crashing through the German defenses in Europe. There were POW interrogations and documentation searches in the overrun rocket launch sites and supply factories. Lists of major personnel involved in the secret weapons program were drawn up very rapidly, aided immensely by the discovery of one document in particular.

The famous Osenberg List was drawn up by an SS professor at the University of Hanover. Goering had placed him in charge of the Reich's Research Council, and the diligent Osenberg had compiled a complete list of research projects and the scientists and engineers assigned to them. The list ran to over fifteen thousand personnel with accompanying Gestapo security checks on each. Embarrassing details of their private lives were often used to keep them under control. As the American tanks of the US 3rd Armored Division approached Bonn in early March, the university there was ordered to destroy all documents relating to weapons research.

Those that were not burned or torn up were flushed down the toilets and it was down one of the toilets that a Polish lab technician managed to salvage some shredded copies of what turned out to be the Osenberg List. Suitably restored, the document was passed on to a British Intelligence agent. From the excited hands of MI6 it was eventually passed on to Major Staver. The list provided him with his most wanted person and location: Wernher von Braun and Nordhausen.

Von Braun had also been instructed to destroy all the documents created by the development of the V-2. With over sixty-five thousand diagrams charting every slight change in the evolution of its design, von Braun was not going to obey the order that would destroy such a wealth of knowledge. Even if Hitler did not see a future, and the Führer was getting desperately near the end of his dictatorship, von Braun saw a time beyond the war when the achievements of his rocket would be acknowledged and the next stage in its development would be based on exactly the documents he was being ordered to destroy now.

Planning for the survival of his rocket team included the safe deposit of the documents, which von Braun recognized as being worth far more than their weight in gold. As false reports of American tanks approaching Mühlhausen, only twelve miles to the south of his position, reached von Braun at Bleicherode, he put into action a plan to hide the documents as a future bargaining tool. He entrusted the burial of his treasure to Bernhard Tessmann and Dieter Huzel, the chief designer of the test facilities at Peenemünde and von Braun's personal aide respectively. He would have done it himself if he hadn't broken his arm in two places and shattered his shoulder in a car crash a couple of weeks before.

It was not only documents that were being burned. Nordhausen and its Central Works, Mittelwerk, became the destination point for concentration camp prisoners evacuated from the east, most notably Auschwitz. Slave labor continued to be required for the construction of the V-2s. The SS had made Dora the main camp of the Nordhausen area, but also established a network of three dozen sub-camps making up the "Concentration Camp Mittelbau."

One particularly horrendous camp was the converted army barracks at Boelcke Kaserne, which became the dumping ground for the worst cases of malnutrition and exhaustion. As more prisoners entered the area, the death rate increased to such a level that the SS crematoria were unable to cope with the bodies. Prisoners were literally worked to death. There were cases of subversion and sabotage among the Nordhausen workforce and the guilty were hanged publicly within the mountain complex. Despite the inhumane environmental working conditions inside the vaults, the documentation that did survive shows that the Mittelwerk produced a total of 5,789 V-2s.

In another example of the horror of war, the RAF made a couple of air raids on Nordhausen, which firebombed the camp at Boelcke Kaserne, killing fifteen hundred prisoners. There was a certain déjà vu from when the RAF had accidentally hit the Trassenheide camp during the original Peenemünde raid. As a result of the raids, the SS started the evacuation of the Mittelbau camp system, forcing nearly thirty thousand inmates onto railroad cars and transporting them to Bergen-Belsen. Those for whom there was no transport were forced to march, and the Nordhausen area was soon littered with dead bodies, left where they fell either due to exhaustion and exposure or a bullet from their guards.

At Gardelegen, the SS herded over a thousand exhausted prisoners into a barn and set fire to it. Fire was being used to destroy evidence of the system that manufactured the V-2s, but even then, there was just too much death to cover up. The figures are only estimates, but it is generally believed that twenty thousand prisoners died within the Mittelbau camp system, out of a total of perhaps sixty thousand. It is also estimated that five thousand people were killed by V-2 attacks on Britain and Continental targets, so one can easily conclude that more people died in the production of the wonder weapon than were killed by its operational use.

Nordhausen was liberated by the Task Forces of the US 3rd Armored Division on the morning of April 11th, 1945. They were not prepared for the horrors that awaited them. Concentration camp survivors did their best to educate them since, as an advance tank unit with no technical intelligence specialists in tow, they were unaware of the significance of Nordhausen and the Mittelwerk. Major William Castille, the Military Intelligence officer with Combat Command "B," had been told to expect "something a little unusual in the Nordhausen area." As he entered the large tunnel on the southern slopes of the Harz mountain, it was like being in "a magician's cave." Originally the subterranean complex had obviously been two railway tunnels penetrating the mountain for more than a mile, which had then been joined together by a large number of cross tunnels. Castille saw both V-1 and V-2 parts arranged in neat rows and the entire vaulted complex looked as though it had been abandoned without any attempt to destroy its contents first. He reported his findings back to Military Intelligence in Paris, which forwarded on the news to Colonel Holger Toftoy, chief of the US Army Ordnance Technical Intelligence.

Up until that point, Toftoy had had no success in acquiring any operational V-2s, let alone a hundred of them, as Colonel Trichel had requested the previous month. Not a single V-2 had been seized at a launch site, mainly because the missiles had been transported in, swiftly erected, and launched. Now, the news of Nordhausen's liberation and the basic report from Military Intelligence indicated that the factory that had produced the missiles was in American hands. Toftoy would be able to fulfill Trichel's seemingly impossible order after all.

To achieve this, Toftoy launched Special Mission V-2, tasking his aide, Major James Hamill, to lead a team of Ordnance experts to Nordhausen and arrange the transfer of the rockets to the Belgian port of Antwerp where they would be shipped to the United States. Hamill had been in charge of the Ordnance Technical Intelligence section for artillery and submarine mines. The new field of ballistic missiles was to become his specialty. At twenty-six, Hamill was already a physics graduate from Fordham University with three years' experience in the US Army Ordnance. He had served with Toftoy in the United States and they would continue to serve together when the V-2s were sent to New Mexico, but that was in the future. First of all the missiles had to be extracted from Nordhausen, a process which would be far from easy and far from legal.

Back in November 1944, the European Advisory Commission, made up of representatives from the US, Britain, and the Soviet Union, had decreed that upon Germany's surrender "all factories, plants, shops, research institutions, laboratories, testing stations, technical data, plans, drawings, and inventions" would be securely maintained until the arrival of the correct occupying power. Because of the strength of the US thrust over the Rhine, the American troops were presently occupying a very large area of Germany that the Yalta conference had agreed was eventually to belong to the Soviet Union. Colonel Toftoy was faced with the problem that the V-2s came under that decree and Nordhausen was located within that allocated Soviet region. But Germany had not formally surrendered yet, so technically he was free to extract the rockets.

"Unofficially," he told Major Hamill, "I'm telling you to see that those V-2s get to Antwerp. Remove all the material that you can, without making it too obvious that we've looted the place."

There was another problem with the extraction. This one did not concern legality but involved the breaking of a gentleman's agreement. If two V-2s were captured, one was supposed to go to the United States and the other would be given over to the British. If only one V-2 was acquired, Toftoy would have to hand it over to the British, since their research facilities were nearer. With this handshake agreement in place, Toftoy would have to obtain two hundred operational V-2s to satisfy Colonel Trichel's order. There was no way that the US Army Ordnance would part with even a single V-2, so Toftoy warned Major Hamill to keep the Special Mission V-2 a secret from the British.

Ordnance Technical Intelligence Team 1 consisted of Hamill, Major William Bromley, and Dr Louis Woodruff, a civilian advisor and professor of electrical engineering at the Massachusetts Institute of Technology. When they arrived in Nordhausen, the first problem presented itself. There were no fully operational V-2s. The underground complex and surrounding area did not possess a store of ready made missiles. If Trichel's order of a hundred V-2s was to be fulfilled, the Army Ordnance team was going to have to collect a hundred of each component that the factory possessed and assemble the missiles back in the United States. Not knowing the exact construction of a V-2, how could they be assured of collecting the full list of components? It became important to find some of the rocket specialists or at least documents listing the components.

Major Bromley, as the member in charge of the technical operations, realized that the physical process of removing the missile parts out of the tunnel and transporting them out of the Nordhausen area was going to be a far lengthier task than originally thought. The Allied bombing that had softened up Nordhausen before the ground troops moved in had destroyed all but one of the railroad bridges. Bromley found that several hundred German rail cars had been stranded because of this logjam and they could be used to transport the V-2s out of the future Russian zone, if only he could move the components down from the mountain. Needing a workforce with a basic knowledge of and respect for machinery, Bromley put in a request for Army Ordnance's own 144th Motor Vehicle Assembly Company. Even that came with problems, since the Company was nearly eight hundred miles away in Cherbourg.

Major General K.B. Wolfe arrived in Nordhausen on part of a fact-finding tour of Europe for the US Joint War-Navy Committee studying German secret weapons for the Foreign Economic Administration. This was one of the organizations supporting the CIOS. It was a fact-finding mission about fact finding which found very few facts in either London or Paris. Scientific intelligence gathering was not being given the priority that CIOS had originally been created to deal with. Combat operations were dominating the attention of the various intelligence services. Wolfe found great resistance to his pleas for help from General Lucius Clay, deputy commander of the Group Control Council. There were far too many intelligence agents active in the theater, and in General Clay's personal opinion, the whole business of scientific exploitation was at least six months premature. There were far more important concerns. He even suggested that study groups such as Wolfe's should not be sent again to Europe in the near future.

Wolfe was fully aware that the Allied forces were presently occupying areas of Germany which they would eventually have to give up to the Russians under the Yalta agreements. Scientific intelligence could not be put off for six months because the opportunity would be lost forever. General Clay had to juggle the twin planning problems of ending the war and starting the peace. He was not impressed either by Wolfe or his case. General Wolfe eventually reported that despite General Clay's reluctance to make scientific intelligence a priority, there was a pressing need to stop the wanton destruction of laboratories and factories by the recently liberated workers, and more importantly to prevent the dispersal of key personnel.

Some American aircraft companies had already let Wolfe know that they would consider taking on the captured German experts as employees. This would achieve the double strike of preventing German rearmament in the near future, and greatly increasing the speed with which American industry could catch up with German technology. In fact General Wolfe was the first to suggest that the German scientists should be actively recruited to work in the United States. As he made his report to the Joint War-Navy Committee and the Foreign Economic Administration, he found few who would support such a suggestion.

Wild Bill Donovan, the head of the OSS, had tried to convince President Roosevelt that it was in the United States' best interests to recruit members of the former German security services who were offering their knowledge against the Soviets. Roosevelt had been adamant that he was never going to allow such people to become residents and eventual citizens of the United States. Donovan's choice of recruits numbered men from the SS, the Gestapo, and the Abwehr, whom Roosevelt viewed as probable candidates for war crimes prosecutions. However, Roosevelt's untimely death and Truman's succession as president removed an important barrier to the transfer of human talent to the US to complement the transfer of technology. Still, the okay to import German scientists was not given straight away.

On April 20th, 1945, Major Robert Staver flew from London to Paris on hearing of the fall of Nordhausen. The CIOS Black List prime target of Nordhausen had now become available, but Staver soon found that access to it was heavily restricted. Until it was classified as safe, G-2, the Military Intelligence, refused to allow any non-combatant into the area. It did not want any of the myriad intelligence services interfering in the war zone, especially those with troublesome civilian scientists tagging along. Although Staver was a major in the Army Ordnance, he was seconded to the CIOS and teamed up with the civilian General Electric Project HERMES team led by Dr Richard Porter. It was his job to guide the HERMES group to the V-2 documents and key rocket personnel. As they waited in Paris for clearance to go into Nordhausen, Staver sent some of the team off to areas which had already been passed as secure. Dr Porter visited the University of Heidelberg and the Technical University at Darmstadt where he interrogated scientists involved in the field of guided missiles.

As the end of April approached, and Staver was still stuck in Paris, he approached Colonel Toftoy for help. As a member of the new Rocket Branch, Staver was based within the Research and Development section of the Technical Division of Army Ordnance. His commanding officer was the highly sceptical Colonel Quinn in London, who thought the whole German rocket business a load of "Buck Rogers junk." Colonel Toftoy, on the other hand, was chief of the rival Intelligence section of the same Technical Division. He already had a team in Nordhausen—Major Hamill's Special Mission V-2—and Staver felt that the search for documents and scientists should obviously run concurrently with their exploitation of the hardware. Toftoy produced fake papers for Staver and one of the HERMES group to be assigned to his Ordnance Technical Intelligence Team 1, with the valid passes for entering an area still off-limits to non-combatants.

Before he left for Paris, Staver received some CIOS information, obtained by British Intelligence, that placed von Braun and other leading Nazi rocket specialists in a village outside Garmisch-Partenkirchen, hundreds of miles away from Nordhausen in the Bavarian Alps. With Nordhausen and the possibility of Central Work documents now open to him, Staver chose to go there, his No.1 location target, rather than go after his No.1 human target: Wernher von

Braun. The Bavarian Alps were not yet in American hands. However, if events progressed at a fast rate—and recent evidence suggested that, as the war approached its end, they might—and von Braun came within reach, then Staver authorized Dr Porter to abandon all other Black List personnel and go for the big one.

Accompanied by HERMES member Ed Hull, the same man who had witnessed the V-2 landing on the Rolls-Royce factory while on his way to Farnborough, Staver made his way to Nordhausen, arriving at more or less the same time as another team of experts. These were not scientists, but lawyers sent to Nordhausen to investigate the war crimes at Dora and the rest of the Mittelbau Camp complex. In a few cases, Staver and Hull would be looking for the same individuals as the team of lawyers.

Major Bromley was an old friend of Staver: the two had graduated together from Stanford back in 1940. In his search for rocket specialists in the area, Staver also found some missile parts for his colleague. With the help of second lieutenant M.S. Hochmuth, Staver found an almost complete Schmetterling anti-aircraft missile in an artificial cave complex near Wofleberg, two miles northwest of the factory. Also hidden there were the guidance and control unit of an Hs298 missile. Other finds included phototheodolites, a range of optical tracking instruments in a pigsty, and a complete guidance unit for a WASSERFALL hidden in a barn. V-2 parts were found in local woods but Staver was not so lucky in hunting down scientists.

Meanwhile, Major Hamill, in charge of Special Mission V-2 at his base in Fulda, sixty miles southwest of Nordhausen, was having trouble finding other members of the US Army to help him in his secret task. The Transportation Corps, who could normally be relied on to provide rail transport, refused to help. They saw the whole V-2 extraction affair as illegal and likely to blow up in their faces. Hamill was left with only one alternative: to run his own railroad. Plenty of unemployed German railway men eagerly offered their services. All Hamill had to do was get the V-2 components down from the factory to the railhead, which would be possible when the Ordnance's own motor transport company, still on their way from

Cherbourg, finally arrived. To add more pressure to the schedule, it was now rumored that the Russians would be taking over the Nordhausen area on June 1st, less than one month away.

Hamill also used former concentration camp workers to assist in the clearing out of the factory. The work involved heavy manual lifting and the packing of components, but the idea of returning to the factory with its horrific memories did not deter one hundred and fifty of them from taking up employment. The ventilation system in the tunnels had been damaged during the initial search of the complex by the American soldiers, as a direct result of which the workers were limited to only eight hours a day in the tunnels. With such a race against time, one wonders what working hours Major Hamill would have imposed on them if environmental conditions had allowed.

On May 12th, 1945, Staver received an intelligence tip on the whereabouts of the former general manager of the *Elektromechanische Werke*, Karl Otto Fleischer. Through this man, Staver started to build up a network of key personnel still in the area, including probably the second most important member of the Peenemünde team, Eberhard Rees, the former director of Prototype Production. Together with a third rocket specialist, Walter Riedel, whom Staver had rescued from the beatings of his American interrogators over a supposed bacteriological bomb, but only after Riedel had lost all his front teeth, the Germans provided the names of more and more contacts. Not all of the names given were keen to help the Americans. One in particular had heated political arguments with both Staver and Hochmuth. This German had no intention of advancing American knowledge of the V-2. His name was Helmut Gröttrup.

News reached Staver that von Braun and a large contingent had surrendered themselves and were being held in Garmisch-Partenkirchen. When, on May 18th, Dr Howard Robertson, a civilian assigned to the Field Intelligence Agency, Technical (FIAT), another one of those myriad Allied intelligence units General Clay had moaned about, turned up at Nordhausen and demanded that

Fleischer, Rees, and Riedel be transferred to Garmisch-Partenkirchen, Staver refused to let them go. He had not yet found any of the V-2 documentation. If anything, it should be von Braun and his men who should be brought to Nordhausen. Dr Robertson, a Cal Tech physicist and scientific advisor to General Eisenhower, failing to pick up Fleischer, Rees, and Riedel, nevertheless passed on information to Staver which finally led to the successful acquisition of those V-2 documents which had so far eluded him.

Von Ploetz, General Kammler's intelligence officer, had revealed under interrogation that Dornberger told another general the documents were hidden at Kaliwerke in Bleicherode, walled up inside one of the mine shafts there. Staver decided to attempt to trick Fleischer into revealing any information he may be keeping from him. Using the details provided by Dr Robertson, Staver implied that the German officers mentioned in the report had firmly identified Fleischer as a co-conspirator in the plot to hide V-2 documents. Staver left Fleischer to stew over the matter, pretending to have no real interest in the validity of the accusation, while actually hoping that he would be upset that such statements had come from his former superiors.

On the same day, the 144th Motor Vehicle Assembly Company, under Captain E.W. Mandeville finally arrived in town after a 177-mile journey from the port of Antwerp. The V-2 components could now be transported down from the mountain factory to the railhead at Nordhausen, where the German rail cars were waiting to take them out of the future Russian zone of occupation. That was the plan, but on the following day, May 19th, Hamill discovered that all two hundred of the rail cars had been impounded by the US Transportation Corps. Although unwilling to assist in the unlawful extraction of rocket parts, the Transportation Corps agreed to move the empty rail cars out of the area and into the American zone. No amount of reasoning on Hamill's part could alter the fate of his cars.

That was until the sole surviving rail bridge out of Nordhausen was dynamited during the night. The circumstances surrounding the destruction of the only route out for the impounded rail cars were investigated by the Counter-Intelligence Corps (CIC) and the finger of blame was pointed at a group known as the "Werewolves guerrillas," though others placed suspicion firmly on Hamill and his Ordnance staff. The destruction of the bridge bought time which allowed Hamill to convince the Transportation Corps of the supreme importance of Special Mission V-2. By the time the 186th Combat Engineer Battalion had repaired the bridge and built another one, the first train of V-2 components was ready to leave Nordhausen. Driven by German railway personnel and guarded by the soldiers of Company B, the train took its valuable load as far as Erfurt where men of the US Army Military Railway System took over, accompanying it the rest of the way to Antwerp.

Fleischer had considered the pros and cons of maintaining his silence on the whereabouts of the V-2 documents and came to the conclusion that his relationship with the Americans was now more important than any loyalty he had towards Nazi officers who had evidently sung like birds during their interrogations. With suitable embarrassment, Fleischer admitted to Staver that he had withheld the most important piece of information from him, but in order to make amends, he would go off with Rees and try to locate the documents. Staver showed great trust in these two Germans, bearing in mind the great deception they had played on him for the past few weeks, but it paid off when they returned after a trip to the specific mine at Dörnten and negotiated with the local guardian of the treasure, Herr Nebelung.

The documents needed excavating from the mine, so Staver sent off Second Lieutenant Hochmuth to deal with that while he himself caught the first flight out of the nearby airfield. Hitching a ride in a single-seat P-38, Staver made the cramped journey to Paris in two hours and went straight to the Army Ordnance headquarters in the Plaza Athénée Hotel. Colonel Toftoy was briefed on the latest developments at Nordhausen and Dörnten, while Colonel Joel Holmes, the Chief of the Technical Division and Toftoy's immediate superior, agreed that priority should be given to the extraction of the documents and their safe transport into the American zone as soon as possible. He immediately assigned two ten-ton semi-trailers for the task and issued orders that they rendezvous with Major Staver at Dörnten.

The other reason for Staver's sudden flight to Paris was that he needed to draft an important cable to the Chief of Ordnance at the Pentagon, Washington. In it, he suggested they seriously consider the recruitment of German rocket specialists within the United States. Their knowledge on ballistic missiles, V-2 and, more specifically, the anti-aircraft WASSERFALL would be of great use to the US Army in the Pacific War, he said. He stressed that it was the considered opinion of the scientific advisors in his team that the German technology was twenty-five years ahead of that of the United States, and that, given three months in the US, the rocket research team promised to deliver the reconstructed drawings and designs of the WASSERFALL. Obviously, the original documents would speed up the process hugely. Staver was suggesting that the top members of the Peenemünde research and development team, numbering one hundred, be transferred en masse to the US. The Germans had already expressed a preference for working for the Americans, with the British second and the French third. Staver stressed that a firm statement of policy on recruitment would be required from Washington in order to not lose these human assets to one of the other Allied Powers. Colonel Holmes approved the recommendations contained within the cable, and eventually it was signed by General Eisenhower himself. A key statement made in the cable went on to have profound relevance in the post-war future: "Suggest policy and procedure be established for evacuation of other German personnel whose future scientific importance outweighs their present war guilt."

While Staver was pushing the case for active recruitment in Paris, Hochmuth was supervising the excavation of the mine at Dörnten. Fleischer had already hired three shifts of local miners to work round the clock clearing the mineshaft of rubble. The documents were hidden in two small rooms at the end of a tunnel nearly one thousand feet away from the entrance. On May 23rd, work was suddenly called to a halt by the arrival of a British scientific inves-tigation team on the scene: CIOS Team No. 163, led by Colonel William Cook. Dörnten was to be taken over by the British from May 28th, five days hence, and they were doing the rounds of their immi-nent occupation zone, "getting acquainted with the neighborhood." Hochmuth, not wearing US Army uniform, masqueraded as a German, and he and Fleischer claimed to be German miners carry-ing out a geological survey. As Cook stood by watching them, they had to fill the boxes not with documents but samples of low-grade iron ore which the local miners had helped to suddenly produce.

Cook had also taken his CIOS team to Nordhausen where several boxes of V-2 components were to be sent back to England. The crates were placed onto a lorry by none other than General Electric and Project HERMES' Edwin Hull, who promptly drove off with them. "We never saw that lorry again," complained Cook's deputy, John Elstub.

On May 26th, another Allied team turned up at Nordhausen, but this time it was a Soviet one, consisting of army officers. Staver, who had returned from Paris, was powerless to stop them taking a quick look inside the Mittelwerk factory. By that time most of the V-2 components were crated up and waiting down at the Nordhausen railhead. The Russians were rumored to be setting up loudspeakers in the area and actively trying to recruit German sci-entists to join the Soviet rocket research. Staver reported this aggressive competition from the Russians to Colonel Quinn, his ever-cynical superior who stated bluntly that he didn't care if they were successful.

Nordhausen was still under American control at the time, but Peenemünde wasn't. It had fallen to Soviet troops earlier in the month and Major General Andrey Sokolov had only now just led his scientific team in to investigate the former rocket site. There was practically nothing left of it, so effective had been the combined efforts of Allied bombings and German self-destruction. There were no personnel in the surrounding area with sufficient expertise to assist the Russians in the slightest. The key Peenemünde personnel had evacuated the site long ago.

All that had been left behind were the technicians in the man-ufacturing shops and plants. It would be a constant, depressing state of affairs for the Soviet investigators and it would be repeated at Nordhausen. Stalin was reportedly angry at the meager pickings. Having conquered the Germans and been first into Berlin and Peenemünde, but not to have caught a single rocket scientist of worth was humiliating. When Eisenhower, as President of the United States in 1957, declared to the press that the Soviets had taken all of the German scientists from Peenemünde, and it was because of this that the Soviet Union beat America into space, one can now see how great a lie that was.

Two items of interest were found at Peenemünde by the Russians, and both related to the influence of Russian rocketry on Werner von Braun and his team of designers. One was the heavily annotated German translation of Tsiolkovsky's book on rockets and space flight and the other was a classified design of a Soviet missile by Mikhail Tikhonravov, obviously stolen from NII-3 in the late 1930s.

Only one of the two semi-trailers made it all the way to Nordhausen, the other broke down en route, but Staver was able to obtain six two-and-a-half-ton trucks from the 71st Ordnance Heavy Maintenance Company to make up the lost capacity. There was then a frantic last minute race over the mountains to the Dörnten mine, where fourteen tons of valuable documentation was loaded onto the trucks. The illegal convoy managed to avoid the legal British roadblocks by just two hours.

In the rush to evacuate the V-2s at Nordhausen, the components had been neither packed properly nor waterproofed for the ocean voyage ahead. This caused problems at the Belgian docks, where the crates were held back from being loaded onto ships. As the number of trains bringing in V-2 components increased over the next nine days, attempts were made to correctly pack and water-proof them at the docks. This hectic activity attracted the attention of British Intelligence agents in Antwerp, who quickly reported that a very large shipment of V-2 missile parts was in the process of being loaded onto Liberty ships. Major Hamill supervised the Belgian stevedores unloading the 341 rail cars, waving aside any complaints about priority since he now had a card issued from the Supreme Headquarters Allied Expeditionary Forces, signed by General Eisenhower himself.

As the sixteen Liberty ships were taking on board their valuable cargo, British efforts were underway to stop the shipment. British officers on General Eisenhower's staff made the strongest protests to the head of SHAEF, and he eventually, reluctantly, decided to block the shipment. However, the delay in the decision allowed Colonel Joel Holmes to give the order for the Liberty ships to sail from Antwerp and by the last day of May 1945, the members of Special Mission V-2 were able to report that the objective from Colonel Trichel had been successfully achieved. One hundred V-2s were on the high seas, heading for the United States.

The documents so hastily excavated from the Dörnten mine and denied to the British were now also safely under American lock and key in Paris. The estimated worth of those documents, in saved research and development time and money, had made the whole reason for Staver's mission to Europe worthwhile, and then some. Their value was put at between $400 and $500 million.

The Surrender and Dispersal

When SS General Kammler heard that the American troops were approaching the Nordhausen area, he had decided on a desperate plan which would endanger the Peenemünde elite. Of the estimated five thousand rocket specialists and their families who lived in the Mittelwerk area, Kammler would select the top five hundred scientists and engineers and transfer them four hundred miles to the south, to the Bavarian Alps, where the last stand would take place. Kammler planned on using them as a bargaining tool for his own safety. For that reason, the five hundred would be guarded by the Sicherheits Dienst (SD), the security service of the SS, and if the bargaining failed, they were to be shot, thus denying the Allies the fruit of German technological research. Obviously, von Braun and his men had to be told a different story—that they were to be transferred to the Alpine Redoubt where the Führer and his SS divisions would regroup. A special SS camp would be made available to them at Oberammergau to continue their weapons research, they were told, but the suddenness of the move meant that their families and relatives would have to travel at a later time.

On April 3rd, 1945, as the five hundred Peenemünders boarded the Vergeltungs-Express, the "Vengeance Express" as Kammler's private train was called, von Braun waved goodbye to them with his one good arm, the other still in a heavy plaster cast from the car crash in mid-March. He was given permission by Kammler to travel by car, on a journey that must have brought back terrible memories. The accident had happened on the autobahn from Weissenfels to Berlin in the early hours of March 16th. Von Braun was being driven to the Ministry of Munitions to discuss the transformation of the Leuchtenberg fortress into a valve factory, and hoped to reach Berlin before daybreak. Night-time was the safest time to travel on the autobahn, but von Braun's civilian driver still had to use dimmed lights for fear of Allied planes. Von Braun took the opportunity to doze off. Unfortunately, so did his driver.

Their car swerved off the road and flew over a railway embankment, crashing down into the sidings. Von Braun remembered waking up during the short flight, then pulling his driver from the wreck just before it burst into flames. He then noticed his severely broken arm and lost consciousness. Luckily for them, another pair of Peenemünders had been following them on the autobahn. Bernhard Tessmann and the chief architect Hannes Lührsen performed first aid on the two men, then spent the next four hours trying to find an ambulance. Lührsen, it must be remembered, was one of the men rounded up with von Braun and jailed by the SS. Tessmann would later be given the task of hiding the V-2 documents in the mine at Dörnten. Von Braun recovered in hospital, as did his driver, who suffered a fractured skull, but the huge plaster cast on his chest and left arm showed the severity of the damage. His arm would never fully heal.

As von Braun arrived at Oberammergau on April 4th, 1945, he found the *Elektromechanische Werke* engineers already billeted in a former Alpine regiment camp with charming views of the snow-capped Alps and Ammer Valley, but surrounded by a barbed wire fence and the armed guards of the SD. The continuation of research seemed possible since the Peenemünde wind tunnel, which had survived the RAF raid, had been dismantled and reconstructed at nearby Kochel. Huzel and Tessmann arrived at Oberammergau from their document-hiding mission, with the news that for security reasons, only one man in Nordhausen knew its whereabouts, and that was Karl Otto Fleischer.

Fearing that Kammler's plan was to kill the engineers, von Braun did his utmost to at least get them removed from the heavily guarded camp and dispersed around surrounding villages. This he achieved by pointing out that the camp was an easy target for Allied bombers. As long as the Reich was still alive in the Alpine redoubt they were to be kept alive, but at the first sign of Allied entry into the mountain fortress, von Braun had no doubt that they would all be executed to stop them and their knowledge falling into the hands of the Americans, British, or French. As the five hundred engineers were spread over twenty-five local villages, their status as hostages was confirmed by the continuation of the SD "protection." Von Braun, in continual pain, soon found himself back in hospital for the re-breaking and setting of his arm and shoulder.

On April 25th, as French troops were reported to be closing in on his hospital in Sonthofen, von Braun believed that his day of execution had arrived when a German soldier entered his room. But the intruder was not SS. He had been sent by none other than General Dornberger. The surgeon was called to apply a new plaster cast to his arm and chest, then von Braun was rushed out of the hospital into a waiting ambulance which took him to the Alpine holiday resort of Oberjoch. There, at the Haus Ingeborg hotel, he was greeted by General Dornberger, his brother Magnus von Braun, and several other Peenemünders. Scattered about were a few SD soldiers but not in the numbers present at Oberammergau. Dornberger's men outnumbered them and, if the need arose, could overpower them, but as time went on and the SD commandos felt deserted by Kammler, they eventually abandoned their distinctive uniforms for those of the normal Wehrmacht.

As the Peenemünders occupied their time at the hotel by playing chess and discussing the future not of missiles but rockets into space, the war in the outside world was coming to a very swift end. On May 1st, 1945, the engineers were listening to a German State Radio Service broadcast of Bruckner's Seventh Symphony in E Major, when the transmission was interrupted by the announcement of Hitler's death that afternoon, "fighting to the last breath against Bolshevism." Hitler had in fact committed suicide in his bunker deep beneath the Reich Chancellery in Berlin the day before.

Expecting the next announcement to be the formal surrender of Germany, Dornberger and von Braun were shocked to hear that the fight was to continue under the leadership of Hitler's appointee, Admiral Dönitz. With the Americans already in control of Oberammergau, and many of their former colleagues already in what they deemed to be safe hands, the two men decided to take the initiative. They would seek out the American forces and surrender under their own terms. Magnus von Braun was chosen to be the emissary, since he spoke very good English.

And so it was that Magnus cycled his way down from the mountain resort of Oberjoch to the hamlet of Schattwald where he made contact with the advancing soldiers of the Antitank Company, 324th Infantry Regiment of the 44th Infantry Division. Private First Class Frederick P. Schneikert was guarding the approach to the village on the Austrian border and ordered the young, well-dressed German off his bicycle and to put his hands up. Over the next thirty minutes, Magnus tried to convince him in both English and German that his brother, Wernher von Braun, the inventor of the V-2, was hiding nearby and wanted to surrender with the rest of his design team. It was imperative that they be taken "to see Ike as soon as possible."

Schneikert was not able to arrange a meeting with General Eisenhower, but he did the next best thing, which was to take Magnus to the local command post of the 44th Division's Counter-Intelligence Corps at the village of Reutte. There, the young German warned First Lieutenant Charles L. Stewart that his brother and the rest of the team were in great danger from the SS who, acting under General Kammler's orders, would sooner execute the team than allow them to fall into the hands of the Americans. Stewart sent him back to Oberjoch with instructions on how to surrender themselves safely, then passed on information to headquarters that von Braun had been found.

Wernher von Braun's brother Magnus returned to the Haus Ingeburg in the early afternoon to inform everybody of the successful encounter with the former enemy. Dornberger and von Braun selected the initial group that would surrender to the

Americans, deciding to keep the soldiers of Dornberger's staff at Oberjoch until the scientists and engineers had safely made the transition into American occupied territory. Magnus had failed to mention to Stewart that there were more soldiers than civilians in the "team." Tessmann and Huzel were invited along in the first party because they knew where the all-important V-2 documents were hidden and these would be the major bargaining tool in the negotiations with the Americans. Von Braun and Dornberger wanted to keep the Peenemünders together as a team and that would require re-assembling the employees who, as far as they knew, were dispersed all over Germany.

Magnus led the convoy to Schattwald, where his safe-conduct pass allowed them to be escorted to Reutte, a town now in darkness, thanks to the destruction of the electricity supply. Lieutenant Stewart received them in the candlelit mansion which now served as the CIC headquarters. Dornberger was surprised to find several of his former staff waiting for him. The Americans had already started assembling the Peenemünder group for him. As the new arrivals' identities were confirmed, they were led to their rooms just as the electricity supply was restored. They were allowed to freshen up, and then were taken to a dining room where they were treated to a meal better than any they had tasted in Germany for the last few years. Von Braun later reminisced: "They didn't kick me in the teeth or anything. They just fried me some eggs."

Fresh eggs, white bread, butter, and coffee. The staple diet for the American fighting man was a veritable feast for the Peenemünders. Suitably fed, von Braun chatted with Lieutenant Stewart about the state of German missile technology. He horrified his host with stories of a rocket in development that would have reached New York, but there were also fantastical plans to send man into space. The German genius stressed that his and his team's real interest was in space travel, not missile development, but that war had usurped their peaceful intentions.

After von Braun and the others retired to their bedrooms on the upper floors of the mansion, Stewart sat alone in his office composing his first impressions of the German rocket specialists. As he set about filling in his report he heard a noise on the staircase outside his room. It was followed by a second sound—one which Stewart instantly recognized as being that of a weapon's safety catch being released. Rushing to investigate, he left behind his own .45 automatic pistol, so had to challenge the mysterious intruder as he climbed the stairs. "Where are you going?" he shouted out.

"To kill those German swine!" came the drunken reply. Stewart recognized the heavy accent as belonging to one of the more temperamental Polish refugees who had cooked eggs for the Nazi guests. Fortunately, Stewart was able to disarm the man otherwise the future of the American space program would have died in their sleep that night in Reutte.

The next morning there was a press conference and photographs were taken of the group. Dornberger and the others, dressed in long leather coats, were outshone by the charismatic Wernher von Braun, made even more striking by the huge arm and chest cast which projected his bent left arm out in a half-hearted Nazi salute. He knew that all the attention he received could only strengthen the team's chances of obtaining work contracts in the United States. Such was his confidence that he made a private wager with the US soldier who had first ordered Magnus off his bicycle in Schattwald, Private Schneikert, that he would be in the United States before the American could return.

After the press conference, Lieutenant Stewart wondered what to do next with his prize guests. Seeking advice from Seventh Army Headquarters, he was told to interrogate them to see which ones were ardent Nazis. Stewart knew the worth of the rocket specialists and knew that Nazi allegiances were probably easily explained as necessary for their own safety during the Reich, but he was after a firmer directive from his superiors. They too were uncertain what to do. The indecision stretched all the way to Washington. General Eisenhower sought clarification on whether the scientists should be kept at special detention camps for a set period or indefinitely.

When the decision was made, it was only a temporary one. The

Germans were to be sent back into Germany, to Peiting. There they joined with more Peenemünders rounded up from further afield. With security at a bare minimum, they were free to wander around the small village and wonder exactly what their future would hold. It would take five days before the Americans finally got their act together, herded the Germans into trucks, and drove them to Garmisch-Partenkirchen where Dr Richard Porter was waiting to interrogate them.

Porter was thirty-two years old with a doctorate in electrical engineering from Yale. As head of the General Electric Project HERMES team sent to Europe, he was viewed as the most qualified American scientist to understand and exploit the advanced technology of the Germans. Though Major Staver had left him instructions in Paris to drop everything if von Braun was captured, Porter had actually heard about the surrender while he was in London on VE Day. Wasting no time, he joined the rest of the HERMES team, officially registered as CIOS Team 183, and headed for Garmisch-Partenkirchen, where an interrogation center was being set up. Porter soon found that other scientific investigators were gravitating towards the former site of the 1936 Winter Olympics. Unlike cases elsewhere, the British were on the ball with this one.

Based at the foot of the highest mountain in Germany, the Zugspitze, the interrogation center was a former military administration building now surrounded by barbed wire and American guards. At this point the document cache at the Dörnten mine was still hidden, so the only scientific intelligence available was expert debriefing of the prisoners. The Germans were willing to talk about their recent prowess, but soon found themselves repeating basic information to different groups of interrogators. CIOS was once again living up to its alternative name of CHAOS.

Dr Porter was aware of the fact that the answers to questions were remarkably consistent and similar, as if the prisoners had been briefed on exactly what they were to say and what they could divulge. The ringleaders were Dornberger and von Braun, and they were playing a game with their interrogators—withholding information until the conditions of future employment in the United States could be negotiated from a stronger position.

For the benefit of Dr Porter, von Braun prepared a lengthy document, with a similarly lengthy title *"Ubersicht" uber die bischerige Entwicklung der Flussigskeitsrakete in Deutschland und deren Zukunftsaussichten,* which translated into "Survey" of Previous Liquid Rocket Development in Germany and Future Prospects. In it, he likened the development of the rocket to that of the aircraft, with huge sums of money being required to produce the same level of advancement as that seen in aircraft from World War I to World War II. Setbacks would undoubtedly occur, but with financial resolve, the rocket could be used for civilian as well as military projects. One such civilian use would be the exploration of space. "When the art of rockets is developed further, it will be possible to go to the other planets, first of all to the Moon. The scientific importance of such trips is obvious. In this connection, we see possibilities in the combination of the work done all over the world in connection with the harnessing of atomic energy together with the development of rockets, the consequences of which cannot yet be predicted."

General Dornberger was equally prophetic in his report: "Scientific high-altitude rockets, a station in space, travel to the moon and to the stars...That state will be first in space which has the courage to make a clear decision. The stratospheric travel rocket will come as certainly as the modern locomotive followed the Stevenson locomotive."

These comments on space travel and the future developments of rockets were meant to convince the interrogators that the Germans were peace-loving scientists who had been made to create weapons of war against their wishes. Porter and his team knew that von Braun and the team were civilians and as such could not be held guilty of war crimes, but General Dornberger could be. Yet he was the head of the group and worked strict discipline on the Peenemünders. When the crucial question of what had happened to the V-2 documents came up, Dornberger claimed ignorance beyond the fact that General Kammler had taken them and hidden them in a mine near Bleicherode. This was exactly the same story told by von Braun. Tessmann and Huzel, the men responsible for the burial of the documents, kept quiet.

As the interrogations wound down, most of the five hundred detainees at Garmish-Partenkirchen were released to start the painful search for their families. Others had climbed over the fence, but the two hundred that remained weren't planning on leaving until they had got an employment contract. Lieutenant Colonel John O'Mara, the Army Air Force officer in command of the camp, and the leading American on the wasted Sanders Mission to Moscow and Blizna, provided the Peenemünders with technical lectures and a library. They formed an orchestra and performed theatrical productions to alleviate the boredom.

On June 1st, 1945, the expected Soviet occupation of the Nordhausen area did not take place as planned and as feared by the Allies. The rush to evacuate the Mittelwerk of as many V-2 components as possible had succeeded in fulfilling Colonel Trichel's order for a hundred V-2s, but the total exploitation of Nordhausen was not achieved. The official reason for the Soviet delay in occupation was given as the occurrence of a very important parade in Moscow which required the presence of Marshal Zhukov and the other high-ranking Soviet military leaders serving in Germany. A new date was set for June 21st, 1945.

Having succeeded in stripping Nordhausen of V-2 rockets, the Americans now set about stripping the area of rocket personnel too. The extra time until the take-over date allowed Staver to request the presence of von Braun and Dornberger to aid the search. What he got was Ernst Steinhoff and Martin Schilling. Together with Rees and Riedel, Staver asked them to give him some idea of how many people would be required to recreate a missile development organization. Their estimates ranged from 350 to 750 and Staver knew that his mission to evacuate such a large number of specialists with their families from hiding places he didn't even know of yet, would be akin to the biblical exodus.

As the days counted down, Staver's searchers managed to round up some of the key personnel, but they came up against resistance from those who questioned whether it might not be a better deal to stay and work for the Russians. The offers that had been made to them from Soviet agents sounded far more enticing than those coming from the Americans, mainly because the Americans were not offering anything. No contracts were promised; all the Americans were saying was that the Germans should flee from the murderous Russians. Yet the Russians were offering them homes and research facilities in Germany. Staver needed someone to sell them the American alternative, and that salesman could only be Wernher von Braun.

On June 19th, only two days before the Soviet take-over date, Staver arrived in Garmisch-Partenkirchen and took matters into his own hands. He grabbed hold of von Braun and Dr Porter and drove them by jeep to Munich from where they were flown to Nordhausen. Three hundred vehicles were hastily assembled, each with a German on board ready to pass on the fact that von Braun had returned and was pressing for an exodus. These Germans were sent off into the surrounding area with names and addresses. As the specialists and their families were brought into Nordhausen, with only twenty-four hours to go, they waited anxiously at the railhead. Like the V-2 transport problems three weeks before, the railway was causing headaches for Ordnance. The rail cars were there but the locomotive engine was not. As an alternative convoy of trucks was being considered, the engine finally arrived. Over a thousand German rocket specialists and their families, carrying all the possessions they could grab in the fifteen minutes they had been given, boarded fifty rail cars and the train left Nordhausen. On board was Helmut Gröttrup, finally persuaded to flee from the Russians. He watched as the armed American soldiers prevented the non-specialists from getting on to the train.

Despite the excavation of the V-2 documents from the Dörnten mine, other important papers still remained in the area. General Dornberger had announced that his own papers were hidden somewhere near Bad Sachsa and it was up to Staver and Dr Porter to locate the documents before the Soviets moved in. As with all treasure hunting, it helps to have a map, and Staver and Porter went to extraordinary lengths to obtain a large-scale map of Germany. They drove sixty miles from Nordhausen to the 332nd Engineer Regiment in Kassel, expecting to find a map there. They were shocked to find that they didn't possess one either. Armed with a mine detector, a small search party was sent out from Kassel with only a few hours left and, with a piece of luck at last, found a large-scale map in a German Forest Bureau. Dornberger's five metal-lined boxes were buried under the ground and the mine detector successfully found them. With another demonstration of last-minute action, the US Army had managed to acquire important documents and simultaneously deny them to the Russians.

Since the beginning of the month, General Eisenhower had been held by the European Advisory Commission decree he had signed which stipulated that German military installations had to be "held intact and in good condition" for the incoming Russian forces. That decree applied to Nordhausen, though Dr Porter was one of those who regretted not blowing the place up when they had finished looting it. As it turned out, the Soviets once again postponed their take-over of Nordhausen until July 1st, but with a thousand of the Germans and their families safely inside the American zone at the village of Wizenhausen, there were no more last minute missions.

On June 28th, 1945, Army Ordnance Colonel John A. Keck announced the capture and interrogation of the rocket scientists. At a Paris press conference he informed the newsmen of the more fantastical projects conceived by the twelve hundred experts, including the mirror in space which could redirect a concentrated beam of sunlight down on anywhere in the world and burn it to a crisp, and the submarine launching of missiles. Keck expressed relief that Hitler's men had not been allowed to develop these futuristic weapons at a faster rate, but commented favorably on their willingness to abandon the national cause in the name of science, by volunteering to continue their research in the United States and Britain. Keck was jumping the gun, since the contracts were still being drawn up in Washington and would prove to be radically different from what von Braun and his men had been promised.

Colonel Toftoy was no longer based in Paris; he had been promoted to take over Colonel Trichel's post at the Pentagon. As the new Chief of the Ordnance Rocket Branch, he would be in a stronger position to deal with any problems over the transfer of Germans. There was understandably a lot of resistance to allowing former Nazi scientists into the country. The War Department General Staff had to carefully regulate who could be accepted. The guidelines it laid down would hopefully prevent any war criminals slipping in and the contracts were set for a limited time. The whole program for the exploitation of Germans was to be handled by the Military Intelligence Service of the War Department and given the code-name Project OVERCAST. The scientists selected would be those for whom there was a definite case that their work was indispensable and could not be carried out in Europe. They would be offered six-month contracts after which, if their work was finished, they would be returned to Europe. Their families would remain in Germany under American protection in American-provided accommodation.

The OVERCAST guidelines were forwarded on to General Eisenhower with the total number of scientists allowed set at 350, exactly the same number as Staver and Dr Porter had negotiated with von Braun. The problem was that 350 was the total number for all sectors. The rocket specialists were to take up no more than one hundred. To supervise the selection, Colonel Toftoy was sent back to Europe. His mission would be a difficult one because the American deal was no longer the best one going.

The Soviet forces had taken over Nordhausen in mid-July and Major Chertok, a thirty-three-year-old guidance system engineer who had previously been a member of the RAKETA group at NII-1, found the Mittelwerk plant in a far better state than the Peenemünde site he had just left. Despite the looting by the Americans, there were still a few V-2s in various stages of assembly. Setting up headquarters in Bleicherode in the same villa that had served as von Braun's home, Chertok and his team managed to quickly gather up about two hundred German technicians who had not accepted the American offer of a train ride out of Thuringia. Their mission was twofold: to collect as much intelligence on the V-2 as possible and to re-establish the production lines at the Mittelwerk factory. It was re-named the Institute Rabe, short for "Raketenbau und Entwicklung," the German for rocket manufacture and development. It had another meaning, for "Rabe" also meant "raven" in German. As a sign of the joint Soviet-German nature of the project, Chertok shared the leadership of the Institute with a local engineer named Gunther Rozenplenter.

In nearby Lehesten, more than fifty brand new combustion chambers were found in an underground depot and they would soon be used to restart the static testing at that location. As the Soviets started to build up their intelligence operations, the Americans appeared to be doing the reverse. The major organizations that had been involved in the hunt for V-2s and their designers were disbanded. The Combined Intelligence Objectives Subcommittee (CIOS) teams were dissolved along with the Supreme Headquarters of the Allied Expeditionary Forces.

As Stalin learned of the plundered state of Nordhausen, he chose the conference at Potsdam to make a complaint to the new US president, Harry S. Truman. The Soviet Union was standing firm on the Yalta demands for a $10-billion reparation from Germany in the form of industrial plant and labor. Part of that should have been the V-2 components, documents, and specialists. Marshall Zhukov provided a list of plundered equipment and personnel from thirty-nine cities in Saxony and Thuringia. When Stalin informed Truman that the US Army had removed a massive number of unauthorized assets from the Soviet occupation zone, the president claimed that whatever was missing would be accounted for and that he was not to worry. As for snatching personnel, Truman denied that the US Army had removed anyone. American forces already had more than enough Germans to worry about, he explained to Stalin.

When Colonel Toftoy arrived in Witzenhausen, von Braun and the Germans were already aware of the contracts being taken up by technicians in Nordhausen. The Russians were providing them with double the rations that the Germans were receiving and rates of pay were good. The Institute Rabe had expanded to take in three centers of research: the Mittelwerk, Bleicherode, and the original site at Peenemünde. New, longer-range rockets were being planned by the Russians, and efforts were being made to coax von Braun and Steinhoff over to the Soviet zone. All that Toftoy could offer was a six-month contract for one hundred of the Peenemünders. Von Braun had wanted a three-year contract for himself and his team, with their families, in the United States. The Americans had to show that they were serious about developing rockets.

Toftoy knew that the issue of families was one he would not be able to match the Russians on. They had the advantage of keeping production in its original German locations, unlike the massive relocation of the American plan. The best that Toftoy could offer was the setting up of a secure compound at Landshut where the families would be accommodated, fed, and protected during the specialists' absence. The salaries earned in America would be paid direct to the families in Landshut, with the men subsisting on $6 a day. For some, no amount of money was enough to entice them to leave their families. One of these was Helmut Gröttrup. He told Toftoy that he did not want to work for the Americans if it meant leaving his wife and two young children. He would rather give up being a rocket engineer and earn his living as a scrap metal dealer, which is exactly what he did in Witzenhausen for a short while. In August he was approached by the Burgomeister of Bleicherode who had crossed the River Werra which separated the American and Russian zones of occupation. He was offered a job at the Institute Rabe with the promise of a top salary and better accommodation than he had previously had. Gröttrup made two secret trips across the Werra to Bleicherode where he met Major Chertok and negotiated his contract. By mid-September he would be back in Bleicherode, with his family, on a salary of five thousand marks ($1250) per month—more than the three thousand marks being offered to von Braun by the Americans.

PART II: THEIR GERMANS AND OUR GERMANS

The Spoils of War

As the Americans were busy looting Nordhausen and the sur-
rounding area, the British 21st Army Group had been having some
success stumbling across abandoned rockets in the Netherlands.
They also managed to capture some of the missile launching crews.
Combining them with the designers who had surrendered to the
US Seventh Army seemed like the obvious way to fully understand
the V-2. The idea was originally proposed by an aide to Major
General Alexander Cameron, the chief of the Air Defense Division
(ADD) of SHAEF. Junior Commander Joan Bernard's idea was to
arrange a fully documented demonstration of the V-2s being
launched while the techniques were still fresh in their prisoners'
minds. The idea was sold to the British Imperial General Staff by
the assistant chief of ADD, Colonel W.S. Carter, and then to the
Supreme Commander of SHAEF, General Dwight D. Eisenhower.
Colonel Carter named the operation BACKFIRE, but the overall
responsibility was given to General Cameron.

A site was selected for the launchings in the old Krupp naval
gun range located a few miles south of Cuxhaven on the North Sea
coast. It had ready facilities for the transportation of heavy equipment and the firing range could conveniently allow radar tracking
of the missiles during flight. ADD search parties were scouring
Germany, France, Holland, and Belgium for V-2 components, covering nearly three hundred thousand miles in six weeks, and finding
enough to fill two hundred trucks and four hundred freight cars.
Five trains arrived from Nordhausen with 640 tons of special tools,

but not a single rocket from the US Ordnance booty. BACKFIRE was definitely a British operation and the Americans were not keen to help.

Two members of the British CIOS Team 183 arrived at Garmisch-Partenkirchen to question General Dornberger about the dangers involved in launching a V-2. He described the safety precautions to be taken with the V-2 propellants and warned them that the weapons should be fired as soon as possible after they had been assembled. Cooperating with his British interrogators, Dornberger provided them with a list of thirty men presently at Garmisch-Partenkirchen who would be of valuable assistance to Operation BACKFIRE. These joined a group that would eventually number eighty-five, selected by the British to participate in the test-launchings at Cuxhaven. During the recruitment period, SHAEF was dissolved and the British component of ADD became the Special Projectile Operations Group (SPOG). The American authorities at Garmisch-Partenkirchen released the required number of rocket experts for BACKFIRE into the hands of the SPOG but only on condition that they were returned as soon as the project was finished. Maintaining high security, the British refused to divulge much information about their requirements to the Germans or even the whereabouts of the test range before transporting them away in six army trucks.

Upon arrival at the Cuxhaven, they were divided into two groups and billeted at Camp A nearby Brockeswalde and Camp C at Altenwalde. The reason for the division was simple. Those at Camp A, under Dr Kurt Debus, would be used for the preparation, check-out, and launching of the V-2s, while those at Camp C would be used to confirm the procedures carried out by Camp A. By this process, the British hoped to eliminate any deliberate sabotage or disinformation from the Germans. Dornberger arrived separately with Major R.T.H. Redpath, the senior intelligence officer at SPOG, but was kept incommunicado from both camps since his power of influence over the German workforce was deemed to be too intimidating. In total, there were about a thousand Germans and two and a half thousand

British military personnel and civilians involved in Operation BACK-FIRE. Many of those billeted at Brockeswalde and Altenwalde were given brand new Nazi Party uniforms to wear since these were the only clothes that the British quartermasters could find. The sight of these men locked behind barbed-wire fences confused the German locals, who believed them to be high-ranking party members who were waiting to be executed. Attempts were even made to rescue them, but the locals were even more confused when they refused to be liberated.

Seventy tons of ninety-three percent pure ethyl alcohol was found near Nordhausen and transported to Altenwalde, where it was diluted with distilled water to produce 80 tons of fuel—enough for twenty V-2s. Since the SPOG efforts at locating missiles had so far failed to produce the desired eight, there was plenty of fuel left over for an illicit distillery to produce schnapps.

In mid-August, Operation BACKFIRE finally obtained the V-2s it needed for the testing. It was not due to the charity of the Americans, but to the discovery of a dozen missiles in excellent condition near Leese, to the west of Hanover. From those, the eight V-2s would be finally prepared for testing. As the BACKFIRE schedule continued with an anticipated launching at the beginning of October, the Americans put pressure on the British to return twenty-six of the Germans to Garmisch-Partenkirchen. The rocket specialists were needed for the war against Japan, an excuse originally given by the US army to the politicians back in Washington. Despite the problems that it would cause BACKFIRE, the British allowed fourteen to return. With the sudden end of the war against Japan announced on September 2nd, the previous justification became redundant and the Americans replaced the BACKFIRE Germans with twenty-five different ones. The fourteen were spirited off to the United States.

With all this bartering of personnel, the main characters had managed to avoid being drawn into the British project until the beginning of September. Dornberger was at Brockeswalde, but incommunicado. Eventually, von Braun, Eberhard Rees, Steinhoff, and Axster were requested to travel to London for a meeting with the British authorities. The Americans agreed to release them begrudgingly for a period of ten days before they were due to fly to the United States. Reuniting them with Dornberger, the British billeted them in a POW camp in Wimbledon and drove them each day to Shell-Mex House in London for meetings with Sir Alwyn Crow, Controller of Projectile Development in the Ministry of Supply. The daily journeys took them through areas of London which had been destroyed by V-2s.

Expecting to be given a hard time by Crow, von Braun was surprised to find that the discussions were amicable and businesslike. Major Staver was also present, forever protecting his prize asset. In addition to asking von Braun and his team whether the larger team assembled at Cuxhaven was capable of doing the job, Crow asked them if they could assist in the creation of a missile development center in England. Staver took this to mean that the British were attempting to steal his prize before his very eyes. Crow even asked them if they would reconsider their contracts with the Americans and stay in Britain. If not, then could they possibly suggest suitably qualified alternative personnel? Dornberger and von Braun said that they thought it would be wiser for Crow to negotiate with individual scientists, since the future of rocket development in Britain would obviously be on a far smaller scale than that likely in either the United States or the Soviet Union.

Staver escorted his team back to Germany, but without Dornberger, whom the British wanted to question further about other matters. He was taken to a POW camp for high-ranking Nazi officers, where he was told to dress in a light brown uniform with the white letters PW (Prisoner of War) on the back. This was to be no normal interrogation about the V-2. He was transferred to the London District Cage, headquarters of the British War Crimes Investigation Unit, at 8 Kensington Palace Gardens. Its commanding officer, Lieutenant Colonel Andrew Scotland, informed him that since the British had failed to capture SS Obergruppenführer Hans Kammler for trial at Nuremberg, they were going to settle for him instead. Someone had to pay for the V-2 assault on London.

Dornberger desperately pleaded his innocence, claiming that he had no control over the firings of the missiles or their targets, that he only produced them. Colonel Scotland was neither swayed by that nor by Dornberger's pointing out that Allied bombings of German cities had killed far more civilians than the V-2s. Dornberger's fate rested with the British Cabinet and Sir Hartley Shawcross, the chief British prosecutor at the war crimes trials at Nuremberg. When Colonel Scotland suggested that he spent the time awaiting trial writing up a full account of the V-2 development and its future possibilities, he refused. Because of that, the British sent him to a castle at Bridgend in South Wales which served as a detention center for high-ranking German officers. It would be his home for the next two years.

Colonel Toftoy was not pleased about losing Dornberger, but the British provided him with tape-recorded conversations between the general and other inmates of detention centers, which clearly showed that he was constantly turning ally against ally and that he would be a source of irritation and future unrest among the German team if he were sent to America. The British considered Dornberger to be a "menace of the first order" who deserved to be "left on the dust heap." As far as they were concerned he was an ardent Nazi who would always yearn for the day when the Reich could be resurrected, and the knowledge gained during Allied patronage could be used to develop superior weapons. During interrogation he had casually mentioned the fact that French, Polish, and Russian prisoners of war had been used as forced laborers at Peenemünde and that many of them had been killed by the RAF bombing raid back in 1943.

While Toftoy agreed with the decision, another officer in another branch of the US armed forces, General Kerr of the Army Air Forces, still wanted to bring Dornberger to the United States, specifically Wright Field air base. His efforts were blocked by another general who would rather "trade him to the Russians for a dish of caviar." However, Dornberger was spared an appearance at Nuremberg when the scale of the atomic bombings at Hiroshima and Nagasaki put the V-2's efforts in the shade. Instead, the British authorities kept him prisoner until 1947, when his release and relocation to Wright Field, just as General Kerr had wished, eventually led to a directorship of Research and Development at the Bell Aircraft Company.

Von Braun and Eberhard Rees were to be in the advance party of V-2 specialists who were to enter the United States under the auspices of Project OVERCAST. Together with five other Germans, they left Frankfurt on September 12th, 1945, in the back of a US Army truck, and headed for the French border. As they crossed the bridge over the river Saar, von Braun said to his dejected companions, "Well, take a good look at Germany, fellows. You may not see it for a long time to come."

August Schulze, a former systems engineer in the Test Laboratory of the *Elektromechanische Werke*, questioned this comment. "What do you mean? You know we only have a six-month contract with the Americans."

"We may have only a six-month contract now; but I still don't think we will be back for a long time to come." It was always von Braun's intention to re-establish Peenemünde in America.

The truck made an overnight stop at a US Army camp near Reims then arrived in the suburbs of Paris early the next morning. The driver, an American lieutenant, took the opportunity to take a detour and visit his brother in a Parisian hospital, but after several hours of searching had to admit he was lost. Von Braun's offer to help navigate was refused since, according to the American, how could a German know his way around Paris? This seemed to ignore the fact that Paris had been occupied by the Germans until relatively recently, but von Braun's knowledge of the French capital originated from earlier times, before the war, when he had visited the city on several occasions during his school holidays.

Asking instead for guidance from a French local, who proceeded to lead them further astray, they eventually arrived at their destination, Le Grand Chesnay. Situated in the forests of a large estate nine miles west of the city near the military academy at St Cyr, the POW camp was their most idyllic yet. Other teams of specialists were there, recruited by various branches of the US armed forces, assisting in the processing of captured documents for shipment to the United States. For five days, von Braun and his Peenemünde team relaxed in the country surroundings, discussing contracts and the future with their fellow scientists as they all waited for transportation to the land of their former enemy.

At last, on the evening of September 18th, a female driver of the US Women's Army Corps took the team and nine other Germans to the US Officers' Club at Orly Field, where they insulted the feelings of a French waitress over dinner, before being rushed on board a C54 cargo plane. They shared the flight to Newcastle Army Air Base in Wilmington, Delaware, with a few US soldiers returning home. The long journey was broken up by several refuelling stops at the Azores and in Newfoundland, but it did not finish in Delaware. The Germans were transferred to a chartered DC-3 which flew them to Logan Field in Boston on the 20th. From there, they were driven in US Army sedans to Boston Harbor and onto a small boat which took them the remaining five miles to Long Island and Fort Strong. This old island fortress had been built just before the end of the last century to protect Boston from a sea-born invasion but now served as a local headquarters for the US Army Intelligence Service.

For the next two weeks, the Peenemünde team were photographed, fingerprinted, and interrogated by intelligence officers asking exactly the same questions as those already dealt with back at Garmish-Partenkirchen. Von Braun was suffering more than most because of the chilly winds irritating his arm and shoulder injuries, and the recently added inconvenience of hepatitis.

On October 1st, Major James Hamill, hero of the plundering Special Mission V-2, arrived at Fort Strong to take charge of the

Peenemünde team. While looking through the US Army Custody Document, before signing it, Hamill left blank the entry "Probable date of return to Fort Strong." It would be a valid answer since none of the team would ever return to that inhospitable place.

Six of the seven rocket specialists were put into the custody of Private First Class Eric Wormser, who took them by train to the Aberdeen Proving Ground, located between Washington and Baltimore, while Hamill escorted von Braun to the Pentagon for important meetings with the bigwigs from Ordnance. Major General Gladeon Barnes, the Chief of the Technical Division, and his chief of Research and Development, Colonel Leslie Simon, informed von Braun of the hundred V-2s that had been successfully extracted from the Soviet zone and of their transfer to the White Sands Proving Ground. Von Braun was to be sent to Fort Bliss, near El Paso, Texas, to begin work as soon as possible.

Hamill had originally hoped to spend some time with his wife, but as he later complained to Colonel Toftoy, he was "made to honeymoon with von Braun." His orders were now to stick to the German "like glue" and not let him have any contact with fellow passengers on the long train journey from Washington's Union Station down to El Paso. At one point, at a transfer in St Louis, they found themselves in the same Pullman car as wounded veterans of the 82nd and 101st Airborne Divisions.

Needless to say, Hamill moved himself and von Braun into another car. Even that led to trouble when a persistent fellow passenger asked von Braun where he was from and what line of business he was in. Von Braun improvised that he was Swiss and in the steel business. His new-found friend was also in steel and had been to Switzerland many times but before the conversation could get too awkward, the train pulled into the other man's stop, Texarkana. As he was about to get off, he shook von Braun's hand and thanked him for his country's help during the war.

"If it wasn't for the help you Swiss gave us, there's no telling who would have won the war."

Von Braun finally arrived at Fort Bliss to receive a very cold reception from the Commanding General there. He was a combat veteran from two World Wars, having been wounded in both, and the German's arrival was totally unexpected. Such was the security surrounding the missile project.

Both Toftoy and Staver attended the BACKFIRE launches. The first, scheduled for October 2nd, 1945, was an embarrassing failure for the British, but the German crews explained that statistically there were always failures. The main propulsion system did not ignite although flames had appeared in the combustion chamber. After extinguishing the fire, the chamber was inspected and repairs made to it but to no avail—a second attempt failed when the pyrotechnic device used to ignite the propellants was expelled from the engine. A second V-2 was prepared for the next day.

On this occasion, the rocket was successfully launched and flew for just under five minutes, travelling the predetermined one hundred and fifty miles in to the North Sea target zone southwest of Ringkoebing, Denmark. It impacted the water only half a mile to the left and one mile short of the bulls-eye. The flight had been monitored by anti-aircraft radars loaned to the British by the Americans. In fact these were the only equipment contributions made by the United States to Operation BACKFIRE.

On October 4th the original test rocket was given a third chance to prove itself, but the launch, though successful, only sent it out a distance of fifteen miles. Despite having eight V-2s assembled, only three were deemed fit for launching. The third and last was scheduled for October 15th when the British had arranged for a special grand finale, code-named Operation CLITTERHOUSE. It was special in that the observers were not just the officers from the British, French, and American armed forces but also those from the Soviet armed forces. The invitation to witness how the Germans launched their V-2s was eagerly accepted by the Russians.

They sent a team of representatives to Cuxhaven, led by General Sokolov who had commanded the forces that had taken over

Peenemünde. Accompanying him were Colonel Yuri Pobedonostsev, chief of the Special Technical Commission in Berlin, and Colonel Valentin Glushko, chief of the V-2 test firings at Lehesten. Two other observers turned up unannounced and this created security problems. Major General Alexander Cameron, still in charge of Operation BACKFIRE despite the disbanding of SHAEF, refused them entry. With a great deal of angry remonstration, they were escorted off the compound and had to view the launchings from afar.

During the pre-launch activities, Colonel Pobedonostsev remarked to Lieutenant Hochmuth that he knew his name and that he had been responsible for removing V-2s and other equipment from the Nordhausen Mittelwerk. He also knew that the V-2s had been shipped over to the United States and were on their way to White Sands Testing Grounds. This was supposed to be classified information, but the colonel was making a point that the Soviet Union knew exactly what had been plundered from their occupied zone. Hochmuth asked him how things were going at the Mittelwerk, and Pobedonostsev replied that they were "having a hell of a time" since the Americans had "cleaned the place out." It was then that the Russian colonel made him an offer. If a Soviet inspection team could be allowed to visit White Sands, they would reciprocate with an offer for a US team to visit Peenemünde. Hochmuth relayed this offer to Toftoy and Staver, but they turned it down. At that stage, White Sands was nothing more than several hundred square miles of desert and a few dilapidated huts, whereas Peenemünde was the bombed out spiritual home of the V-2, more a tourist site than a working research center.

Following the successful demonstration, the observers were taken on a guided tour of the assembly and checkout areas as well as the control bunker. Attempts to get the two uninvited Russians on that tour also failed and they had to remain outside the barbed wire perimeter. One of them, wearing a captain's uniform despite holding the rank of colonel, was Sergei Korolev—the man who would eventually lead the Soviet race into space, and without doubt the most important person in the Russian delegation.

OSOAVIAKHIM

Sergei Korolev was trained in aeronautical engineering at the Kiev Polytechnic Institute and figured in the early Soviet rocket experiments, co-founding the Moscow organization, Group for Studying Reaction Propulsion (GIRD in Russian). Similar in many ways to von Braun's *Verein für Raumschiffahrt* (VfR), GIRD were experimenting with liquid-fuelled rockets of increasing size. The first successful launch in the Soviet Union occurred on August 17th, 1933, when an engine designed by Mikhail Tikhonravov lifted a 40lb rocket off a test stand in the woods of Nakhabino, west of Moscow. The historic flight lasted all of eighteen seconds, but was enough to inspire Korolev to write: "It is necessary also to master and release into the air other types of rockets as soon as possible in order to thoroughly study and attain adequate mastery of reactive techniques. Soviet rockets must conquer space!"

Twelve days after this momentous Soviet achievement, the US Department of the Navy were putting an end to the research work of Robert Goddard, an American pioneer who had beaten Korolev by more than seven years. A letter from the Acting Secretary of the Navy clearly stated that "because of the great expense that would be entailed in development of the rocket principle for ordnance and aircraft propulsion, which under present stringency of funds appears hardly warranted, the Department regrets it is not in a position to further such development."

This fact was unknown to Korolev, since at the time he was spreading the false rumor that both Goddard in America and Oberth in Germany were colonels in their respective armies and carrying out secret rocket developments. Wishing to force his own Ministry of Defense into funding his organization's experiments, all that they seemed to be willing to do was publish a book of his entitled *Rocket Flight in the Stratosphere*.

When Stalin's Purges extended to the army in 1937, it was only a question of time before it would reach the rocket scientists at the Scientific Research Institute 3, known as NII-3. Korolev's boss was Ivan Kleimenov, who had once worked in the Berlin office of the Soviet airline Aeroflot, and that was grounds enough for Stalin to suspect him of being a German agent. His arrest and that of his deputy, Gyorgi Langemak, put suspicion on the whole research institute. Stalin saw conspiracies everywhere and felt it was obvious that the NII-3 was preparing to use rocket weaponry to overthrow his government.

In the early hours of June 27th, 1938, two members of the NKVD and two so-called "witnesses" burst into Korolev's apartment and dragged him away from his wife and three-year-old daughter. Based on the testimonies of Kleimenov and Langemak before their executions, and his fellow worker Valentin Glushko before his sentencing to eight years in prison, Korolev was tortured and confessed to NKVD charges of sabotage. His punishment was more severe than Glushko's—he began a ten-year sentence. Shuttled from prison to prison he was entered into the Siberian Gulag system for the next few years during which he lost all of his teeth and acquired a heart condition.

In 1942, he was transferred to a special prison group in Kazan known as a "sharashka," which acted as a design-bureau-behind-bars, allowing the State to utilize its imprisoned scientific talent while still punishing them for their alleged crimes against the State. By 1944, Korolev had designed and tested a liquid-fuel rocket engine RD-1, which was soon put into production and installed on Soviet fighters and bombers. Stalin rewarded Korolev and the thirty-four other engineers involved in his *sharashka*, which included Valentin Glushko, with their freedom. Having served six of the ten years,

Korolev returned to Moscow to be reunited with his ex-wife—who had divorced him during his period in the Gulag—and daughter.

Major General Gaydukov, the Communist Party's representative for the Special Technical Commission (OTK in Russian), was responsible for all V-2 recovery operations in Germany. He visited Nordhausen in August 1945 to make a personal survey of the work being carried out by his 284-strong team there. He decided that the effort could be greatly enhanced if some of the former members of the GIRD and NII-3 who had been incarcerated under the Purges of the 1930s were transferred to Germany. On drawing up a list of those experts from before the war, Gaydukov noticed that the two main ones, Korolev and Glushko, had already been released.

So it was that in early September, after spending six years in a variety of prisons, held under false charges of sabotaging rocket development, Sergei Korolev was summoned to the Commissariat of Armaments in Moscow where he was awarded the rank of Lieutenant Colonel and employed in the OTK operation to restore the German V-2 facility at Nordhausen. Having spent precious little time with his family after such a long separation, Korolev was sent to Berlin to study the documentation that existed there. It was the only major source of information on the V-2s which was out of reach of the plundering Americans. He was joined there by Glushko, now also a colonel in the Red Army.

After a few weeks in Berlin, familiarizing himself with all the currently available intelligence on the V-2 operations, Korolev travelled to Bleicherode where he met the highest-ranking German to join the Institute Rabe: Helmut Gröttrup.

As one of the few Peenemünde veterans in the Institute, Korolev made full use of his expertise, tasking him to write an engineering history of Peenemünde which would later become "the most complete and objective account of the work at Peenemünde and the technical problems which had to be solved in the course of the first long-range ballistic missile." The history would take him nearly a year to complete. In the meantime, Gröttrup had the additional duty of developing two special trains as mobile launch centers.

Deceptively named the *Fahrbarbare Meteorologische Station* (Mobile Meteorological Station), each train consisted of between eighty and a hundred rail cars offering transport facilities for the V-2s and their launchers, crew quarters, dining cars, and mobile laboratories. These should have been seen as a warning of things to come.

Gröttrup was placed in charge of *Zentralwerke*, a pilot production line in an old V-2 repair depot at Klein Bodungen. From his initial thirty-man "Bureau Gröttrup," Gröttrup soon came to command a workforce at *Zentralwerke* of five hundred. General Gaydukov also made him responsible for all the guided missile development carried out by the Germans within the Soviet zone. In return, Gröttrup was able to provide the Russians with an extensive list of component subcontractors both inside and outside the zone. When certain parts could not be found in the Soviet zone, Gröttrup sent some of his more daring employees across the border into the American and British zones, where they were bartered for, often with Soviet rations of food, drink, or tobacco. Components were smuggled back across but also training manuals for launchings.

Some of the German subcontractors had been based in Czechoslovakia during the war and it was rumored that Czech insurgents had seized a trainload of documents because the fleeing Germans had planned to transport them to a secret burial site. Vasily Mishin, the young engineer from NII-3 who had worked on the V-2s since the days of RAKETA, was sent to Prague to locate the documents. While hunting for paper trails, he discovered that Prague had been the main coordination point of all the V-2 components made in Austria, Hungary, and Poland as well as Czechoslovakia. Having no success locating the train, Mishin even tried his luck with the British administration officials but to no avail. Eventually, with the use of "unorthodox measures" involving the sister of one of the Soviet engineers, the V-2 archives were found but they ended up being very poor versions of the treasure trove found by the Americans.

Nevertheless, bit by bit the Soviet operation in Germany built up its knowledge of the V-2. Korolev worked well with Gröttrup, no doubt aided greatly by the former's fluency in German. The working relationship between Glushko and his team of Germans at the static firing test stands at Lehesten was less harmonious due to Glushko's lack of language skills and hatred of Nazis. It wasn't long before Glushko had moved the Germans elsewhere within the *Zentralwerke* and replaced them with Soviet engineers capable of testing the engines.

In October 1945, the British V-2 launching demonstration at Cuxhaven had been a frustrating time for Korolev. With only three official invitations given to the Soviets, General Sokolov chose Yuri Pobedonostsev and Valentin Glushko to accompany him. Korolev and Lt Col Georgy Tyulin, a veteran of the Soviet Katyusha rocket units, pestered Sokolov to take them along too. They attempted to bluff their way into the British testing facility at Cuxhaven by demoting themselves to the rank of captain, and Korolev even acted as General Sokolov's chauffeur, but the British had refused them entry. Although he had seen nothing of the launching, the visit to Cuxhaven inspired Korolev to reproduce the British experiments with the Institute's V-2s. The test firings were to take place in the Soviet occupation zone of Germany, the operation was code-named VYSTREL, and it was set for early 1946.

At that time, Mishin was back at the Institute Rabe, leading the Calculation Theory Bureau. Korolev asked him to calculate trajectories and develop aiming algorithms for Operation VYSTREL, but concerns about secrecy eventually led to the cancellation of the project.

In Berlin in November 1945, General Gaydukov received a visitor from the Central Committee of the Communist Party. The representative carried a message from the Party leadership that although they were pleased with the progress made in the project to study and reproduce the V-2, the OTK would have to cease operations in Germany within the next few months until a decision was made on its future. Gaydukov was soon to discover that the Soviet armed forces were not particularly keen to produce large ballistic missiles.

Stalin ordered him to select a Soviet ministry that would oversee the operation when it moved to Russia. There were three possibilities for the role: the Commissariat of the Aviation Industry; the Commissariat of Ammunitions; and the Commissariat of Armaments. Gaydukov's first choice was the People's Commissar of the Aviation Industry, Alekey Shakhurin, but his attitude to the future of missiles hadn't changed since he had wound down RAKETA after the V-2 retrieval from Blizna: rocket engines would be better served attached to aircraft. Gaydukov feared that Shakhurin would withdraw most of the engineers that worked in OTK since they were technically still under contract to the aviation sector. Gaydukov's second choice was People's Commissar of Ammunitions, Boris Vannikov, who seemed to show some interest in the job until Stalin intervened and removed him from the list, appointing him to oversee the newly created Soviet atomic bomb project. That left only one choice: the People's Commissar of Armaments, Dimitriy Fedorovich Ustinov. Luckily for Gaydukov, Korolev, and the whole future Soviet space program, Ustinov was exactly the right man for the job.

In February 1946, Gaydukov and Korolev travelled to Moscow to present the case to Central Committee member Malenkov for a more centralized structure to the rocket development effort in Germany. They were successful in replacing the Institute Rabe, which had always specialized in flight-control systems, with a new organization, the Institute Nordhausen, which covered all aspects of V-2 development. It was divided into four plants at four different locations. One of these, Plant No.1, was Mishin's Calculation Theory Bureau, relocated to Sommerda, eighty miles east of Leipzig. Korolev assigned to them the task of fully restoring a set of design documents for the V-2 that Mishin had obtained from Prague. Once it was completed in its original German language, it was to be translated into Russian.

Despite the apparent consolidation, the OTK's performance in Germany was still a problem for the Central Committee of the Communist Party. A special commission was sent to Nordhausen in May 1946, led by the Chief Artillery Directorate commander Marshal Nikolay Yakovlev and the People's Commissar of Armaments, Ustinov. The commission's short visit resulted in a report which altered the future of the Soviet space program and the fate of the German specialists.

On May 13th, 1946, Stalin signed the Council of Ministers' decree "Questions of Reactive Armaments" which established a secret, nine-member Special Committee for Reactive Technology, similar to the one for the Soviet atomic bomb project. One of the key members of the committee was Ivan Serov, the Deputy Minister of Internal Security (MVD). The plan to relocate the V-2 project from Germany to Russia necessitated the acquisition of a suitable factory, and the committee's deputy chairman, Ustinov, found one in Kaliningrad. The M.I. Kalinin Plant No. 88 had been used for manufacturing tanks and artillery weapons during the war, but Ustinov renamed it the Scientific Research Institute No. 88 (known as NII-88, in Russian). Stalin's decree also stated that the work of the OTK would cease in late 1946 with the full transfer of all Soviet and German personnel to Russian territory.

Obviously, this had to be kept secret from the Germans. Gröttrup and the five hundred men under his command had originally been promised by the Russians that they would continue to work in Germany. This condition was one of the major selling points in accepting their contracts over the American ones. However, for the sake of security, Korolev decided to keep the move a secret from most of the Soviet workforce as well. With Serov as a Special Committee member, Korolev knew that the NKVD would not allow mistakes.

In the summer of 1946, Korolev asked Gröttrup and his team to suggest improvements to the design of the V-2 by mid-September. It was a sign that the Russians were starting to consider the next phase in the development of a Soviet version of the V-2. Only about a dozen V-2s had been assembled from the components found in

Germany and these were to be used for training purposes. Gröttrup and his team came up with a list of a hundred and fifty suggestions, of which about half were accepted by the Russians. These included ideas from back in the Peenemünde days, such as the relocation of all control equipment to behind pressurized propellant tanks and the redirection of exhaust gases to drive the turbo pumps for the propellant.

The first Soviet copy was designated the R-1 (Raketa-1) and differed very slightly from the V-2, mainly by having a redesigned tail and instrument compartment. But Korolev was unhappy about the prospect of merely reproducing the V-2 and set about designing the R-2 instead, using some of the improvement ideas from Gröttrup together with some of his own. By stretching the length of the missile by another nine feet and increasing the thrust of the engine from twenty-five tons to thirty-two, Korolev aimed to increase the rocket's range from two hundred miles to four hundred. Instead of reusing the exhaust gases to power the turbo pump, he used his own idea of adding a second pump in series with the first. This would be a sign that the Soviets were starting to seek independence in rocket development. The increase in thrust necessitated a new generation of engine and Glushko's Soviet team were the first to relocate to the Soviet Union, transferring the entire test stand plant from Lehesten to a Special Design Bureau No. 456 (OKB-456) in Khimki, four miles northwest of Moscow.

In early August 1946, members of the Special Committee for Reactive Technology made a return visit to Bleicherode. Ustinov took the opportunity to officially appoint Korolev the new "Chief Designer" of all long-range ballistic missiles at NII-88. Mishin was appointed his deputy and set off immediately to Kaliningrad to serve as acting head until Korolev had finished winding down the operations in Germany. Mishin found the NII-88 facilities very different from the Institute Nordhausen. The former factory was still in a dilapidated state, with a leaking roof and packing crates for tables. The Soviet workforce was billeted in overcrowded barracks and tents, and with the lack of hospital facilities, disease was widespread.

Colonel General Ivan Serov, as the Deputy Commissioner of the Soviet Military Administration in Germany as well as Beria's deputy at the MVD, began planning the mass transfer of German workers and their families. There would be approximately seven thousand in total. Even the MVD would not consider breaking up families the way that the American contracts had. Of those thousands of workers at the Institute Nordhausen, Serov had requested from General Gaydukov a list of the most capable German rocket specialists. He provided 177 names—with their families a total of 495. With the full resources of the MVD, Serov planned to round them all up at once, specialist and worker alike, and put them on trains to Moscow. As the date approached, and Korolev continued to hide the fact from the Germans, he began to question the wisdom and morality of transporting them wholesale to the USSR. "We must have a little more self-respect," he is reported to have said.

To hide the upcoming operation, code-named OSOAVIAKHIM (Obsche Souznoe Obschestvo AVIAtsii KHIMii), Gaydukov continued to make requests of the Gröttrup team. An example was the design of a missile capable of flying a distance of 1,500 miles. Gröttrup merely had to reproduce the Sanger sketches of the proposed A9/A10 which had been discussed back in Peenemünde. Previous suggestions for V-2 improvements were returned for reconsideration. Nevertheless there was still suspicion among the Germans that they were going to be taken to the Soviet Union at some point in the future. The disappearance of the entire Lehesten plant did not bode well. Gröttrup was also aware that the two trains he had helped design were primarily meant for operating outside of Germany. But when the feared day arrived, it still came as a complete surprise, thanks to the ruthless planning of Serov and his secret police.

General Gaydukov arrived in Bleicherode on October 21st, 1946, to hear the presentation of the Germans' recommended improvements of the V-2. Gröttrup found the general to be very receptive to the ideas of his team and the future looked to be moving ahead with eagerly anticipated increases in range and accuracy. After the presentations and discussions, Gaydukov invited Gröttrup and his managers to join him in a party to celebrate Soviet-German comradeship. Gröttrup knew from experience of Gaydukov's previous parties that this would be a very drunken affair. He was right: the standard rounds of toasts to friendship were repeated, aided by a seemingly endless supply of the best vodka specially imported from Moscow.

At about 3:00am Gröttrup's wife, Irmgard, was woken up by a phone call from one of the wives of the engineers, asking her if she was also going to Moscow with the other Germans of the *Zentralwerke*.

"For Heaven's sake! What a time for bad jokes! You must be drunk," she replied before angrily hanging up the phone. It wasn't long before there was a banging at her front door and a Russian soldier was telling her that she didn't have long to gather her things. Outside in the street, and all over Bleicherode and the Nordhausen area, armed soldiers were rounding up the people on the MVD lists. Irmgard managed to phone her husband, who was still at Gaydukov's party.

"Do be sensible!" he warned. "General Gaydukov is with me, the room is full of officers. You understand, don't you? There is nothing I can do. I may come home—on the other hand I may not see you until we get on the train. I may have to fly over in advance or follow afterwards."

Serov's ruthless efficiency achieved the complete rounding up during a few hours of night. On being woken up by armed soldiers, each individual on the list had been handed the following message:

As the works in which you are employed are being transferred to the USSR, you and your entire family will have to be ready to leave for the USSR. You and your family will entrain in passenger coaches. The freight car is available for your household chattels. Soldiers will assist you in loading. You will receive a new contract after your arrival in the USSR. Conditions under the contract will be the same as apply to skilled workers in the

USSR. For the time being, your contract will be to work in the Soviet Union for five years. You will be provided with food and clothing for the journey which you must expect to last three or four weeks.

People were given three hours to pack everything into trucks, then they were taken to the various railway stations in the Soviet zone, where a total of ninety-two trains had been made ready. The enforced deportation of approximately seven thousand Germans would take several days to carry out.

Despite the anticipation of being separated from his wife and children, Gröttrup found Gaydukov to be very accommodating of his wishes. He was taken to the siding at Klein Bodungen where he was reunited with his family and given three compartments of the sleeping cars. On board were fellow specialists identified by Gaydukov in the list he gave Serov. One of them was Dr Ronger, whose wife had recently died. The Russian soldiers who had rounded him up had urged him to grab any woman he fancied, since they could get married in Moscow. This fuelled later wild rumors that all the Germans were urged to take any woman they wanted to start a new life with. In reality, the wives of the engineers were given the choice of accompanying their husbands or staying behind in Germany if their husbands so allowed. There were some cases where the latter choice was made. Unmarried couples were encouraged to travel together rather than be separated.

As the train set off on its long journey to Moscow, Gröttrup dictated a letter of protest to the Soviets. Entitled "Official and Formal Protest Against the Deportation of Central Works Employees to Russia," it stressed the many occasions when both Korolev and Gaydukov had affirmed to him that such action would never even be considered within the next few years. Yet now they were on their way to an enforced five-year exile from Germany and their homes. Relationships which had seemed so cordial at the *Zentralwerke* were now different. Even when they were greeted at Moscow by a familiar face from the Institute Nordhausen—Pobedonostsev. Irmgard

Gröttrup commented in her diary: "We were neither gassed nor received at the Kremlin."

The 177 Germans on Serov's rocketry specialist list were split up into two groups on their arrival in Moscow on October 28th. One group, numbering seventy-three, was sent immediately to the island of Gorodomlya in Lake Seliger, a distance of 150 miles northwest of Moscow. This remote location had been the scene of some of the most intense, bitter fighting between the Soviet and German armies during 1942. Understandably, the local Russian population hated the Germans, and the barbed wire island compound was as much for the Germans' own protection as for maintaining the secrecy of the missile project. Known as NII-88 Branch No.1, the Russians appointed Dr Waldemar Wolff and Josef Blass to lead the German community there.

The other group, of which Gröttrup was the leader, was transferred to a far more hospitable area near Datschen, in the northeastern sector of Moscow itself. This group would be integrated into the nearby NII-88 facility at Kalinigrad. Far from being plunged into communist poverty, Gröttrup continued to be well treated by the Soviets. He and his family were accommodated in a large six-roomed villa, formerly home to a member of the Council of Ministers. Gröttrup's BMW automobile was even transported from Bleicherode, but since the Soviet Union refused to recognize foreign driving licenses, a Russian chauffeur was also provided.

Other members of the German team had to put up with far less luxury. They were based along the Yaroslavskaya railroad, near the stations of Bolshevo, Valentinovka, and Pushkino. Although housed in old Tsarist mansions and vacation houses, the allocation of space was typically communist: one room for a family of three; two rooms for a family of four; and an extra room provided for those who were university graduates. This extra acknowledgement of academic worth was also reflected in their pay scales. The highest paid Germans were those twenty-four people with a PhD, such as Kurt Magnus who received six thousand rubles a month. Those who held engineering degrees were paid four and a half thousand rubles a month. The rest were paid a monthly salary of four thousand rubles, which was still four times higher than the average Soviet rate of pay for an engineer. Mishin was only on two and a half thousand rubles a month as Korolev's deputy, and the Chief Designer himself was on the same scale as Dr Magnus. Gröttrup was paid the most—a massive ten thousand rubles per month, which allowed Frau Gröttrup plenty to spend on shopping, with her chauffeur as escort.

The mass deportation of the Germans had not gone unnoticed by the Americans. In Berlin, Robert Murphy, the Political Advisor for Germany, cabled US Secretary of State Jimmy Byrnes that "though apparently not formally prohibited by any existing agreement these deportations seem to be particularly inhumane." Colonel Frank Howley, the US commandant, protested that the Soviet action was a clear violation of human rights.

The Washington *Times Herald* on October 25th denounced the kidnapping of "a hundred and fifty thousand" skilled technicians. Reuters reported a remarkably accurate figure of seven thousand. The communist newspapers in Berlin tried to set the record straight, pointing out that the Americans had already removed some confirmed Nazis. The critical attitude of the newspapers annoyed the Russians attending a meeting of the Coordinating Committee on October 29th, 1946.

In typical fashion they rejected the press reports as anti-Soviet propaganda and asked why the Americans and British could not have raised the question of the deportations at the meetings instead of using the Western press. The Soviets were claiming that the removal of technicians was legally covered by the Yalta and Potsdam agreements where the personnel were classified as German assets fit for reparation. Though they had complained bitterly about the illegal American removal of the rocket specialists from the Nordhausen area, they had not done so publicly in the press.

General Clay questioned the Russians over whether the Germans had had any choice in the matter or had just been rounded up and deported in a fashion similar to the Nazis' deportation of civilians into the forced labor camps. Clay reminded them that the Nazi Fritz Sauckel had been condemned for such an action at Nuremberg and hanged for his crime. The Russian delegation did not take kindly to the comparison and after these remarks from General Clay refused to discuss the matter any further.

Until the next day. Marshal Vassily Sokolovsky, the chief of the Soviet Military Administration in Berlin, asked Murphy for a private meeting outside the Coordinating Committee, where he made threats. The Soviet Union felt no need to justify its actions to the United States, he said, and if the press campaign continued, the Soviets would retaliate in kind, but at an escalated level. As Sokolovsky put it, poetically, they would retaliate not on the basis of "an eye for an eye, but a jaw for an eye." General Clay's comments the day before could be seen as nothing more than American provocation.

The whole affair showed the hypocritical nature of the United States: everybody knew that they had forcibly removed a large number of Germans from Thuringia ahead of the Soviet takeover of its legitimate zone of occupation. Murphy replied that the Germans he was referring to had not been forcibly removed and that, anyway, the Americans had not questioned the equally large-scale Russian removal of Germans from the same area. Sokolovsky countered that no Germans had been removed before October 21st, 1946. Both sides were making false statements. Sokolovsky terminated the meeting with a repetition of his threat to use the press "blow for blow and give twice and more than it received."

With the Soviets claiming a slightly more justifiable position than the US, and the Americans choosing to not let the matter lie, US Secretary of State Byrnes cabled Berlin ordering Murphy to turn up the pressure at the Allied Control Council in November. More protests were made at the Soviet use of forced labor and the deportation, as examples of its rejection of basic human rights. The Soviet Union again reaffirmed its right to take German scientific personnel as part payment of war reparations. Even the French had questioned the rights of a German to refuse to be part of such reparation.

The American and British delegations were at last silenced by an irate Russian general turning on them with an unanswerable question: "I am not asking the Americans and British at what hour of the day or night they took their technicians," he said. "Why are you so concerned about the hour at which I took mine?"

Back in Moscow, Gröttrup soon realized that some of his former colleagues appeared to be missing. Even allowing for the splitting away of the group that went to Gorodomlya Island, there were still as many as thirty who were not working at NII-88. Fears that their "disappearance" might be a more sinister act by the Russian security services were dispelled when Gröttrup learned that the apparent unity of the Special Technical Commission (OTK) back in the Institute Nordhausen had belied the true nature of the many Soviet ministries that made it up. Now, in Russia, these individual ministries competed for German personnel. Gröttrup had some limited success in regrouping his Nordhausen team back into NII-88.

Twenty-three Germans were sent to work for Valentin Glushko's OKB-456 propulsion development center at Khimki, which was still under the control of the Ministry of Aviation Industry. Among them were designers, engineers, mechanics, shop technicians, and welders, but none had ever worked at Peenemünde. Including their families, this group numbered sixty-five people who lived around Kaliningrad and would ride to work on buses until specially built cottages were constructed nearer the bureau and the missing thirty had been spirited away by rivals. The group's leader was Dr Oswald Putze, who had worked as the technical director of the Linke-Hoffman plant where V-2 combustion chambers had been manufactured during the war. Another notable member was Werner Baum, a former controlling engineer for the V-2 at the Armed Forces Weapons Office, who would later claim that Glushko stole his designs for the combustion chamber that would help put *Sputnik* into space. The hatred that Glushko had felt for the Germans in Lehesten continued at Khimki.

Gröttrup was worried about the legal status of his team in the initial months of their stay in the Soviet Union. Although their letter of deportation had mentioned a five-year stay, they had not yet been issued passports or other identification documents and those with families still in Germany had not been allowed to send letters home. But these problems were nothing compared to the state of their working conditions at the NII-88 plant.

The terrible, disease-ridden factory that had been the home for a large, Russian workforce preparing to receive them and their V-2 components had not developed much beyond dilapidation. The hardware that had arrived from the various plants of the Institute Nordhausen, courtesy of Serov and his MVD, had been unloaded on the snow-covered ground and left to rust out in the open due to a severe lack of storage facilities. The wooden crates which had served as tables for the Russian workforce were still being used as such, and precision tools from Germany had been grabbed by other ministries. After such effort had gone into obtaining the few documents and blueprints in Germany that the Americans had not seized, some of these were then lost in transit. Suspicions were levelled at competing ministries but Gröttrup found no joy from the Minister of Armaments, Ustinov.

On December 6th, 1946, Gröttrup finally received an official reply to the letter of protest which he had written on the train from Bleicherode to Moscow. It stated very clearly that the Soviet Union had exercised its legal right to deport Germans for the reconstruction of the country under the terms of war reparations. If Gröttrup continued to have difficulty coming to terms with that, then he would be sent to the Urals as punishment.

At the end of April 1947, Gröttrup had had enough of the way German creative output was being stifled by the increasingly uncooperative Russians. There was also a lack of the promised contracts. He decided to go on strike and offered his resignation as the leader of the German collective. Instead of the anticipated trip to the Urals,

Gröttrup received improved working conditions and salaries for his men, but at the price of a personal drop in income. His previous ten-thousand-ruble monthly salary was reduced to eight and a half thousand—but this was still two and a half thousand more than Korolev.

During his "strike", Gröttrup was given a special report to read and comment on. It was entitled *"Über Einen Raketenantriebe Fernbomber"* (On the Rocket-propelled Antipodal Bomber) and was dated August 1944. It had been written by Drs Eugen Sänger and Irene Bredt of the *Deutsche Luftfahrtforschung* (German Research Institute for Aviation) in Ainbring. The proposed antipodal bomber was to be boosted into space using a large rocket, then skimmed along the upper atmosphere until it reached the target where it would drop the bomb. This dipping in and out of the earth's atmosphere would, theoretically, allow the bomber to travel great distances and enable New York to be reached from a European launch. The Russian defector Colonel Tokaev would later reveal how great an interest the Kremlin showed in the Sanger project.

Gröttrup read the report and concluded that the idea was full of unrealistic assumptions. For instance, he did not believe that an exhaust velocity of sixteen and a half thousand feet per second and a chamber pressure in the engine of 1,470lb per square inch were attainable. Sanger had not, in his opinion, fully dealt with the problems of re-entry. The wings of the bomber would not be able to withstand the pressures of the skimming phases of the flight. Gröttrup could not see the advantages of a nearly two-mile launch ramp over a vertical take off. Nevertheless, the Kremlin still passed the report onto the other Germans at Gorodomlya Island for their views.

The raison d'etre for the Germans working in the NII-88 was the development of an improved version of the V-2. This would be designated the G1, or German Rocket No.1, and although work started in Kaliningrad in June 1947, the general concepts had originated within the Institute Nordhausen. The G1, however, would have to be redrawn from memory since the Germans were not allowed access to their previous drawings. It was to be 46.5 feet in length and 5.3 feet in diameter, with a gross weight at launching of 40,590lb —much heavier than the V-2 but having a lighter "empty" weight. The new advanced rocket would have a detachable warhead whereas the V-2 warhead had remained attached to the empty rocket body. It was planned that the GI would have an improved propulsion system, producing an increased thrust of 70,400lb. The bootstrap turbo pumps first suggested by Gröttrup's team back in Germany were once again included. The guidance system for the GI consisted of four antennae on the ground establishing a very narrow guide beam in vertical and horizontal planes, which the missile "rode." Its velocity was measured by an on-board Doppler transponder, which was used to determine the moment for an engine shut-off signal to be sent to the missile. The anticipated range of the GI was about 570 miles and its accuracy greater than the V-2s. Korolev ignored most of the German improvements in design, but he showed great interest in the guidance system.

When he resumed full time work again in July, Gröttrup visited the other German community on Gorodomlya Island, where the working and living conditions were worse than Kaliningrad. Any attempt to improve on those would be personally beneficial since it wouldn't be long before the Soviets decided to begin moving all the Germans out of NII-88 and onto the island.

On August 26th, Gröttrup was ordered to make a journey on one of the Mobile Meteorological Station (FMS) trains he had developed back in Germany. He disappeared without getting a chance to say goodbye to his wife. For almost a week, the train took him and several other Germans southwards from Moscow, then eastwards beyond Stalingrad for another seventy-five miles, to a village called Kapustin Yar. The other, military FMS train was waiting for them in a siding of the Ryazano-Uralskaya Railway. They had arrived at the top-secret location of the Soviet Union's first long-range missile site. Originally, it was going to be somewhere else, for the Special Committee for Reactive Technology had settled on the Azov shore.

But when the projected flight routes over the Don steppes towards Stalingrad necessitated the relocation of a large number of Ukrainians, the First Secretary of the Ukrainian Communist Party made a personal complaint to Stalin who agreed to place the launch site elsewhere. That First Secretary was Nikita Khrushchev.

The eight thousand military engineers who had spent the last few months constructing the site were still there, ready to assist in the launching of the first V-2s in the Soviet Union. In one of those strange paradoxes, they were living in green US Army tents. Gröttrup had the luxury of the sleeping car he had helped design, and after October 19th, the company of his wife. Irmgard, fearing that her husband had been abducted again, had pestered NII-88 officials until they eventually gave in and told her where her husband had been taken. She was flown into the dirt airfield a few miles outside Kapustin Yar, and stayed at the secret location for over a month.

The equipment, launching base, and horizontal and vertical static-firing test stands had come from either Peenemünde or Lehesten, as indeed did the machinery left to rust in the sidings. On October 27th, only two days away from the scheduled first launch of the V-2, Gröttrup was talking with a visiting group of officials from NII-88 when, right in front of them, a Russian worker fell 65 feet to his death. With scarcely a break in the conversation, the visitors carried on as the dead worker was dragged off. The next day, another accident occurred close to Gröttrup when a scaffolding beam came loose and crashed to the ground, killing a leader of one of the welding groups. The cold-blooded reaction of the Russians contrasted with that of the Germans: industrial safety precautions had been of paramount importance at Peenemünde.

On the day of the launch, Minister of Defense Armaments Ustinov joined Korolev and Gröttrup at Kapustin Yar. He watched the countdown stop at "zero minus 5" when, to Korolev's embarrassment, the launching platform suddenly collapsed sideways and with it the fully loaded rocket. One leg of the platform had given way due to a broken rivet. Russian workmen rushed towards the platform and, with total disregard for their own safety, winched the whole thing back into position. Propping it up with girders, the countdown was resumed. With a sandblasting roar, the V-2 shot up into the air and headed east, followed by an observation aircraft. Its flight was monitored by cinetheodolite stations along the route. These cine-telescopic cameras, made by the German firm Askania, had come from Peenemünde and were operated by Red Army engineers hastily trained by the head of the German survey section.

Ustinov grabbed hold of Korolev in a bear-hug embrace and danced about. In his turn, Korolev did the same with Gröttrup and then everyone present embraced everyone else. According to Irmgard in a Nova television interview aired on WGBH, Boston, on February 2nd, 1993, "They jumped up and down like little children. High ranking ministers or not, they were just like little children. Then they grabbed their vodka bottles and got drunk." Pandemonium broke out when the missile was reported to have travelled a distance of almost 185 miles and hit the target area. But that was only the first firing; on the next day, the second V-2 was a failure and Ustinov muttered that he suspected sabotage. The reason for this accusation was the scale of the deviation from the target. When the missile veered off to the left early on in its flight and disappeared off the tracking, someone joked that it was heading towards the city of Saratov.

Ivan Serov, deputy chief of the secret police, was present and did not see the funny side. "Can you imagine what will happen if the missile landed in Saratov? I won't even tell you, you can guess yourself, what will happen to all of you."

Gröttrup felt more secure when he found out that Saratov was 170 miles away, at the extreme range for the V-2. Then reports came in that the missile had crashed 112 miles away, and pressure was applied to the Germans to correct it by the next launch. Dr Magnus, the gyroscope expert, earned his top salary by swiftly identifying and rectifying the gyroscope vibration error. It also earned the Germans a huge bonus of fifteen thousand rubles each and a canister of alcohol, when the third day brought another success. But this time Ustinov approached Gröttrup and jeered, "Soviet rockets

better than German rockets after all!" Irmgard had to ask her husband afterwards what the Minister had meant by that comment.
Had the Russians already developed a rocket of their own? Gröttrup
explained that the third V-2 had been assembled from German parts
in Moscow, whereas the previous two had been assembled in
Germany. The fact that the assembly section in Moscow was headed
by a German named Rasper was ignored by Ustinov. Korolev would
not have liked the Minister's comment if he had heard it. He and
Mikhail Tikhonravov were still trying to develop a real Soviet rocket.
Despite the celebrations and rewards, Ustinov ordered Korolev to
investigate the reasons for the failure, since the fault had been suspiciously rectified so quickly. When Ustinov left Kapustin Yar,
Korolev approached the Germans directly, asking them if such
things were common at Peenemünde. Korolev's own admission that
the events were suspicious made Magnus later recall, "That was a
blow below the belt which showed us quickly that we lived in a land
of institutionalized distrust."

In all, eleven of the V-2s were launched before the Gröttrups
were allowed to return to Moscow on December 1st, 1947. Five of
them had reached their targets, the same percentage achieved by
the Germans during the war.

Operation PAPERCLIP

In September 1945, the US Joint Chiefs of Staff (JCS) had established a new organization in Washington to replace CIOS, called the Joint Intelligence Objectives Agency (JIOA). Its European subsidiary was the Field Information Agency, Technical (FIAT), headed by Dr Howard P. Robertson of Cal Tech, whose main aim was to interrogate high-level German scientists at a camp outside Frankfurt, code-named DUSTBIN. Dr Robertson's reading of the JCS 1067 directive was that the German rocket scientists were to be kept in Germany and that any transfer to the United States would be a threat to national security. "In allowing the Peenemünde boys to continue their development we are perpetuating the activities of a group which, if ever allowed to return to Germany or even to communicate to Germany, can in fact contribute to Germany's ability to make war—and it is the avowed principle of the Allied powers to prevent just this from happening." Robertson failed to prevent the transfers, and sarcastically observed that it was because "this group has technical information and abilities which can be used to further weapons development in the States for use against the Japs!" He made that comment several weeks after Victory in Japan Day.

On October 1st, the War Department openly acknowledged its interest in German scientists. It issued a press release announcing, "The Secretary of War has approved a project whereby certain understanding German scientists and technicians are being brought to this country which are deemed vital to our national security." It was stressed that the Germans would be interrogated, and allowed to stay only for a short while under the constant supervision of the War Department. They would not be exposed to any American classified material during their stay.

The first group of Germans had left von Braun at Fort Strong to go to the Foreign Documents Evaluation Center at the Aberdeen Proving Ground, Maryland, where the V-2 documents previously hidden by Huzel and Tessmann in the mine at Dörnten awaited them. There were three and a half thousand reports and over half a million engineering drawings together with the Dornberger files from Bad Sachsa. Private First Class Wormser, holder of a Master of Science in mechanical engineering and a fluent German speaker, had made a mess of organizing the documents before being sent to Fort Strong to get the Germans to make sense of them. Eberhard Rees and his colleagues spent the next few months going through the documents, organizing them by project and subject matter. The team's lack of engineering English necessitated the recruitment of a more technically fluent group of German naval officers from a POW camp at nearby Fort Hunt in Virginia. The language barrier was also unexpectedly evident during recreational trips to Philadelphia, when the team would sometimes stop off in the "Pennsylvania Dutch" region of the state. Here they would meet descendants of the seventeenth-century German settlers who had fled religious persecution in Europe. Communication was difficult with a near three hundred year difference in German language.

The second group of the team arrived at Pier 90 in New York aboard the *Argentina* on November 16th, 1945. On the following day, a *New York Times* reporter disclosed that eighty-eight scientists had disembarked from the ship, shabbily dressed and carrying old patched-up duffel bags. The five US Army Intelligence officers who had accompanied the group from Le Havre had prevented the press from getting close enough for photographs or interviews. A third group arrived in New York aboard the *Florence Nightingale* on December 6th, and joined the previous group at Fort Strong, where they were interrogated and mysteriously exposed to a large number of monotonous films on duck hunting and pheasant shooting. They spent a gloomy Christmas playing Monopoly and learning English.

By the end of December, the first group had finished reorganizing the V-2 documents at Aberdeen. Two Americans from their interrogations in Germany reappeared at the Proving Ground, one with a future in the missile business and the other without. Dr Richard Porter arrived with his Project HERMES team to sift through the documents for anything of use to their contract with the US Army Ordnance Department. The other was Major Staver. After all that he had done for the acquisition of German documents and specialist talent, Staver was being demobilized. Colonel Toftoy had offered Staver's anticipated position at Fort Bliss, Texas, to Major Hamill instead, so Staver was taking the option of retiring with the rank of Lieutenant Colonel in the Reserve.

On January 11th, 1946, the three groups all set off for El Paso, with the final fourth group leaving Landshut, Germany, on the same day. This would take the liberty ship *Central Falls* to New York and had to be segregated from the rest of the passengers who were US soldiers returning home. After interrogation, not at Fort Strong like the others but at Fort Hunt, they would finally arrive at El Paso on February 23rd. The Peenemünde team had been reassembled, but the rocket development site at White Sands was as far from Peenemünde in quality as it was in distance.

Major Hamill was at the head of an organization named the Office of the Chief of Ordnance, Research and Development Service, Suboffice (Rocket). That was about as impressive as it got. Colonel Trichel's original plan had been to set up an American guided missile program, in collaboration with the General Electric Company. Project HERMES was to analyze, test, and understand the advanced German technology then develop American advancements itself. The project had been conceived during wartime and the projected funding reflected that. Now that the wars with both Germany and Japan had finished, and future conflicts seemed unlikely in the near future with the Americans in sole possession of atomic weapons,

Army Ordnance funding had been severely scaled back.

As a result, von Braun and his Peenemünde team were relocated to what they described as "an icebox in the desert." There were no great wind tunnels, no giant test stands, no purpose-built laboratories or tool shops. What facilities there were had been hastily erected from timber made available by the destruction of some old cavalry buildings at Fort Bliss. Even this operation had been an unauthorized one that nearly resulted in the court-martial of the man in charge of preparing the base for the arrival of the Germans, Lt Colonel Harold Turner.

In mid-August 1945, while von Braun and the rocket team were still being interrogated in Garmisch-Partenkirchen, Turner was having to deal with the arrival of the V-2 components so heroically extracted from Nordhausen. The three hundred freight cars were taking up potentially very expensive siding space along a 210-mile stretch of railway from El Paso to Belen, New Mexico. In order to avoid paying demurrage fees for unloaded freight cars, Turner hired local flatbed trucks and, in twenty days, moved the V-2 components to the White Sands Proving Ground, eighty miles from El Paso. When von Braun arrived ahead of the rest of the Germans, he was shocked to find that the V-2s had rusted somewhat from the saltwater in the holds of the leaking liberty ships during the transatlantic voyage. The 111 German rocket specialists recruited under Operation OVERCAST had been designers, not production men or technicians, so their lack of expertise in reassembling the V-2s came as an embarrassment. Luckily, Dr Porter provided the skilled staff from Project HERMES so that the first V-2 was ready to fire at a static test on March 14th, 1946.

A week later, Colonel Toftoy visited White Sands to see how the program he had inherited from Colonel Trichel was shaping up. Major Hamill told him the bad news: of the hundred V-2s looted from Germany, only a third of them could be assembled from the components. Toftoy detailed First Lieutenant Hochmuth to once again look for spare parts, starting in the United States and then going back to Europe if need be. Hochmuth's hunt in America, retracing the route taken by the components from their entry into the United States, was unsuccessful.

So in May he returned to Europe, armed with letters of authorization from the Chief of Ordnance, directing all US Army Ordnance personnel to help him in his quest. After all the American deceit at Antwerp docks and disinterest in Operation BACKFIRE, Hochmuth even had the audacity to ask the British for any spare parts left over from the Cuxhaven demonstrations. Needless to say, the British refused.

Using letters of introduction from some of the German specialists at White Sands, he tried dealing with the German companies that had originally supplied the Peenemünde components. For those businesses still in production, legal restraints had been placed on them by the Allied military government. It was unlikely that parts for German missiles were allowed to be produced. There was also the added problem of getting French firms to admit that they used to provide the Germans with components for the V-2. In the end, Hochmuth's quest was a failure and the missing parts had to be manufactured in America by companies such as General Electric and the Douglas Aircraft Company.

It wasn't just the components that were missing, but also the key tools. When Arthur Rudolph suggested to General Electric that the assembly of the V-2s would be speeded up by the employment of four men on one job instead of one, he was told that only one metric wrench was available. Even an offer to have the machine shop file down some one-inch wrenches that GE would have to purchase from a hardware store in nearby Las Cruces was rejected because of the cost. Rudolph eventually had the wrenches manufactured by some soldier machinists at Fort Bliss.

This example of General Electric's miserly attitude may have reflected the company's relationship with the Germans and not with the US Army Ordnance. But the Germans were merely consultants to American industry and institutions involved in guided missile research. Apart from assisting in the assembly, checkout, and launching of the V-2s transferred from Nordhausen, they were kept apart from any work considered to be an advancement on Peenemünde. This led them into areas where they had no experience, such as ramjet propulsion. An example was the Project HERMES B2 ramjet. The V-2 was to be used as a first-stage rocket for a ramjet-propelled second stage. The idea had been discussed briefly back in Garmisch-Partenkirchen during a period when the Germans had been twiddling their thumbs between interrogations. The fact that the Americans were already developing a ramjet weapon elsewhere was kept from von Braun and his team.

The first V-2 was launched from White Sands on April 16th, 1946, six months after the BACKFIRE demonstration in Cuxhaven and eighteen months before Gröttrup's launch at Kapustin Yar. The rocket reached an altitude of only three and a half miles before one of its fins fell off, necessitating its destruction. The next firing, on May 10th, was a success, reaching an altitude of seventy-one miles, in front of visiting Pentagon officials, journalists, and industrialists. Colonel Turner rewarded his staff and members of von Braun's team with small wooden models of the V-2. One wonders whether they had been carved from the same timber that had nearly led to his court-martial.

Since the Germans were feeling so under-used by the US Army, von Braun wondered whether they had made the right decision to come to work in America. He later observed, "Frankly, we were disappointed with what we found in this country during our first year or so. At Peenemünde, we'd been coddled. Here they were counting pennies. The armed forces were being demobilized and everybody wanted military expenditures curtailed." This was reinforced during a visit to Fort Bliss by Major Staver just before he left the US Army. He reported back to the Research and Development Service in the Pentagon that the rocket team should be given a more challenging assignment than "helping with the firing of a few V-2s in the New Mexico desert."

One of the Germans risked being sent back to Germany in his attempt to get some kind of value put on their presence in America. Dr Ernst Geissler threatened to leave if the Germans were not allowed access to the swimming pool at Fort Bliss. General Homer, the commanding officer at Fort Bliss who had threatened to court-martial Colonel Turner for ripping up his old cavalry buildings, capitulated and offered them its exclusive use one day a week and also similar terms with the base's bowling alley. Once a month the Germans were escorted into El Paso in groups of four, with an enlisted man in tow, for shopping and eating out in restaurants. The local businesses frequently had to deal with $100 bills for small purchases, since this was how the US Army paid the Germans their per diems.

Over in Germany, their families were receiving paychecks and mail from America. The husbands were being paid a tax-free salary on a bi-weekly basis in Germany to a dependent or bank named by the employee. The War Department had stipulated that the amount could not exceed $10 per day for 312 days per year, payable in marks at an exchange rate of ten marks to the dollar. Basing the individual salaries on 1944 tax reports where available, the United States Forces, European Theater (USFET) ended up overpaying the specialists since the exchange rate of the American-issued mark was four times that of the regular mark in the American zone. The living conditions of the Landshut families were also higher than those of Germans living in the surrounding area and this soon led to local belligerence that had to be dealt with by the American civil affairs officers.

One area where the Landshut occupants suffered more was in the delivery of mail. Because of the double censorship—at both the United States and German ends—the letters from their husbands could take as long as ten weeks to reach Piflaserweg 22a V, Landshut, or "Camp Overcast" as the residents and locals indiscreetly called it. With such an embarrassing breach of security, Army Ordnance officers decided to change the name of the recruitment operation from OVERCAST to PAPERCLIP, in memory of the procedure used for selection back in the days of Witzenhausen, when the files of those personnel required by the United States would be identified by an attached paperclip.

With the change in name came a period of legal changes for the German scientist recruitment program. The lack of visas for the original OVERCAST recruits continued to pose problems for the Joint Intelligence Objectives Agency, but the new recruits would be provided with proper immigration documentation to satisfy the State Department. The JIOA had acquired a new role as an official sponsor of those Germans still wanted by the War and Navy Departments. It was to run detailed security checks on those individuals, using its own intelligence files and those of the FBI. The names approved by JIOA would then be forwarded to the Justice Department for final clearance. Once that was achieved, visas would be issued by the State Department. The JIOA Governing Committee was led by a US Navy captain, Bosquet Wev, and met on a weekly basis at the Pentagon. Other members were G-2 Exploitation Branch chief Lieutenant Colonel Monroe Hagood and, occasionally, Colonel Toftoy. The State Department representative was Samuel Klaus, a character who would soon figure at the center of a major conspiracy plot.

Samuel Klaus was a Jew and a Hebrew scholar who had seen at first hand the evidence of the Nazi war atrocities in Europe. As a graduate of Columbia Law School, he was recognized by fellow lawyers as having "one of the finest investigative minds in or out of government." With his bowler hat, cane, and moustache, children mistook him for Charlie Chaplin, but there was nothing comical about his job during the war.

Klaus established the State Department's SAFEHAVEN Project, tracing the intricate network of front organizations and money-funnelling operations carried out by German banks and corporations trying to smuggle their assets out of Germany before the collapse of the Reich. Klaus also helped to set up the Berlin Document Center which housed all the captured Nazi files, including the infamous SS records. On his return to Washington, Klaus felt confident that he understood the realities of the Nazi state better than most.

He was also very suspicious of the German rocket scientists and the War Department's claim that they were outstanding scientists whose presence in America was vital to national security. At best, he believed that their knowledge could be extracted within the six months mentioned in their contracts and that they should then be returned to Germany. Experience from the setting up of the Berlin Document Center led him to believe that some of the rocket scientists were likely to be ardent Nazis who had been allowed to enter the United States without proper immigration controls because of their alleged importance to national security.

Klaus' view was going against that of his superiors who were arranging for the Germans to stay longer and become US citizens. President Truman and Secretaries Byrnes and Patterson feared that the Germans would pose a greater danger to the United States if they were forced to return to Europe, where they were likely to be seized by the Russians. This represented a major policy shift that saw the advantages of denying the enemy over any moral issues.

Klaus, forced to accept the decision of his President, decided to make the application procedures for German scientists wishing to enter the United States as thorough as possible. He was trying to successfully screen out those undesirable Nazis, but the other members of the JIOA just saw his actions as attempts to block the whole process. As a result, Committee Chairman Wev and Lieutenant Colonel Hagood decided to take Klaus out of the loop, denying him access to the JIOA intelligence files and its decision making process. According to Herbert Cummings, a colleague of Klaus, he was accosted on New York Avenue in Washington by a JIOA staffer, Major Simpson, who hissed at him, "Get that little Jew off the committee."

On June 24th, 1946, Major Hamill was requested by the Pentagon to ask for Wernher von Braun's opinion on the capability of Helmut Gröttrup and the Germans in the Soviet zone, especially in the field of developing an intercontinental missile. Von Braun replied that Gröttrup was a very capable leader and that the people known to be working with him were also capable of advancing the existing projects of the A-9, A-10, and A-11 first conceived at Peenemünde. They also had the advantage of having test stands and a complete production plant.

In response to the Russian forced deportation of German scientists in October 1946, some of the Western press learned from "authoritative" sources that the United States had a similar program of its own named Operation PAPERCLIP and that it intended to import up to a thousand German scientists and offer them US citizenship.

From early December 1946 the major newspapers and weekly magazines printed articles on Operation PAPERCLIP, explaining the official government policy, the numbers involved, and the plans for citizenship. The financial savings in research and development costs to the US rocket program were put at a minimum of $750 million. *Newsweek* ran interviews with Ernst Steinhoff, Rudolph Hermann, and Martin Schilling from the rocket team at Fort Bliss. A Gallup poll of December 11th, 1946 asked the question: "It has been suggested that we bring over to America one thousand German scientists who used to work for the Nazis and have them work with our own scientists on scientific problems. Do you think this is a good idea or bad idea?"

The respondents answered in a ratio of ten to seven that it was a bad idea, believing that the Germans were still Nazis and could not be trusted, that they would gain knowledge from working in the United States and use it against the country someday. Those in favor justified their choice by citing German leadership in science and their preference for them working for America rather than against it with the Russians.

The year ended with the press release of a "profound concern" from a group of forty distinguished individuals including Richard Neuberger, Norman Vincent Peale, and Albert Einstein. The text of the release was from telegrams they had sent to President Truman and Secretaries Byrnes and Patterson:

> We hold these individuals to be potentially dangerous carriers
> of racial and religious hatred. Their former eminence as Nazi
> Party members and supporters raises issues of their fitness to
> become American citizens or hold key positions in American

industrial, scientific, and educational institutions. If it is deemed imperative to utilize these individuals in this country we earnestly petition you to make sure they will not be granted permanent residence or citizenship in the United States with the opportunity which that would afford of inculcating those anti-democratic doctrines which seek to undermine and destroy our national unity.

With von Braun and his team at the US Army base at Fort Bliss, Texas, one would assume that the future of American space travel resided there, but there was another service which had its eyes on the stars: the US Navy. As early as October 1945, the Navy Bureau of Aeronautics set up the Committee for Evaluating the Feasibility of Space Rocketry (CEFSR) which performed calculations and profile analysis of a single stage earth satellite vehicle using liquid hydrogen-oxygen as fuel. The Committee's proposal strongly recommended the construction and launching of an earth satellite vehicle to carry electronic equipment for scientific testing. The Navy Bureau of Aeronautics soon found that the estimated $8 million preliminary design phase would not be fully financed by the navy, so sought partners elsewhere. Members of the CEFSR approached the army air force in the hope of setting up a joint services satellite project. When the two services met initially on March 7th, 1946, the army air force representatives showed a great willingness to get involved. They would pass the details of the project on to General Curtis LeMay.

In mid-March General LeMay announced that the army air force would not support the US Navy satellite plan, and that put an end to any hope of a joint project. The inherent unwillingness of military services to cooperate with each other was proved once again with this rejection by LeMay. He immediately set about creating the army air force's own satellite program by hiring the Douglas Aircraft Company in El Segundo to study the possibilities. This led to a joint Douglas-AAF venture known as Project RAND. It issued its first report, completed in just three weeks, in May 1946 entitled,

Preliminary Design of an Experimental World-Circling Spaceship.

The speed with which the report was produced was due to the army air force's scheduled presentation before the next monthly Aeronautical Board of the War Department. The AAF was to present its alternative to the navy design in a battle to secure funding for the development of weapons to be used in the medium of space. The RAND report was amazingly prophetic, as is shown by these excerpts from its introduction:

> Although the crystal ball is cloudy, two things seem clear:
> 1. A satellite vehicle with appropriate instrumentation can be expected to be one of the most potent scientific tools of the Twentieth Century.
> 2. The achievement of a satellite craft by the United States would inflame the imagination of mankind, and would probably produce repercussions in the world comparable to the explosion of the atomic bomb.
>
> Since mastery of the elements is a reliable index of material progress, the nation which first makes significant achievements in space travel will be acknowledged as the world leader in both military and scientific techniques. To visualize the impact on the world, one can imagine the consternation and admiration that would be felt here if the US were to discover suddenly that some other nation had already put up a successful satellite.

The RAND report itself admitted that it was a feasibility study for a satellite vehicle "judiciously based on German experience with the V-2." It believed that technology and experience had now reached a point where it would be possible to construct a craft that could penetrate the atmosphere and achieve sufficient velocity to establish an orbit of the earth. A multi-stage rocket would be required, against the proposal of the US Navy single stage, and the vehicle would need to reach a speed of 17,000mph and be aimed properly to achieve a perfect balance of centrifugal force and gravity. The time frame for the establishment of a 500lb satellite placed into a three-hundred-mile high orbit was five years, i.e. by 1951. Project RAND ruled out its use as a military weapon, since at that time there was no propulsive system capable of lifting a heavy A-bomb to orbital altitude. Lighter explosive warheads were not economically justifiable. The satellite's military uses would be restricted to its surveillance capabilities, as a high altitude observation "aircraft." However, the research and development of a satellite vehicle would have military advantages in that it would lead to the space-travelling requirements of an intercontinental ballistic missile. The Aeronautical Board of the War Department decided on the evidence presented to it by both the Navy and the AAF, that both projects should be independently pursued.

RAND issued a special follow up report in October 1946 entitled *The Time Factor in the Satellite Program*, written by Dr James Lipp. He too had access to a crystal ball:

> The possibility of constructing a satellite has been well publicized both here and in Germany and the data of the Germans are available to various possible enemies of the United States. Thus, from a competitive point of view, the decision to carry through a satellite development is a matter of timing, depending upon whether this country can afford to wait an appreciable length of time before launching definite activity.

On December 17th, 1946, at midnight, a V-2 with several potential satellites on board took off from White Sands. In the warhead compartment was a number of US Army rifle grenades filled with steel pellets. At a height of a hundred miles, the grenades were to explode, firing the pellets further into the upper atmosphere and producing manmade meteorites visible from earth. The Germans and their American colleagues were mystified when nothing was found of the warhead compartment that should have parachuted safely to the ground. Also, there were no sightings of the artificial meteorites. Questions were asked about whether the warhead compartment or the pellets had actually gone into earth orbit. The rocket team ruled out the former idea but considered the latter a possibility.

December also brought a relaxing of the restrictions on families at Landshut. At last they would be able to join their husbands in America. Perhaps the presence of women at Fort Bliss affected those members of the team who were still single. One of them may have been Wernher von Braun, who suddenly travelled to Europe on February 11th, 1947. Army Ordnance officers had to seek permission for him to return to Germany, since his OVERCAST/PAPERCLIP contract only allowed him to do this when "a state of emergency exists." The state of emergency was the need to marry his cousin, Maria Louise von Quistorp, who at eighteen years old was nearly half his age. With US Army intelligence officers present, von Braun enjoyed a short honeymoon then took his young bride back to America. Security had been tight throughout the trip as there were great fears that the Soviets would try to grab him.

The special status of the German rocket scientists showed itself when the FBI reported a suspicious incident involving von Braun's brother, Magnus. FBI agents in El Paso had learned that Magnus had sold a platinum bar to a local jeweler, and were curious as to how and from where he had obtained this. When the incident was reported to the Justice Department in Washington, the FBI agents were told to drop the case. The US Army ordered, "action not be taken for security reasons and possible adverse publicity which might affect the long range objectives of the project on which the group of Germans was employed." The War Department was getting increasingly protective of its German assets as the Nuremberg trials approached.

At the February 27th meeting of the JIOA Governing Committee, JIOA Director Colonel Thomas Ford sat at the table holding a file containing the list of all the PAPERCLIP scientists already resident in the United States. Klaus had spent the past year trying to get that list of names. Ford also had a list of Germans that the JIOA wanted to move through the pipeline who needed visas. He demanded that Klaus sign a waiver giving entry visas to those new recruits. When Klaus asked to see the list, Ford refused to show him, even though the file was there in view. The list was classified and the State Department was not allowed to see it. Ford threatened Klaus if he didn't sign there and then. The information would be given to several senators who would "take care of the State Department."

Director Ford had very good reasons to keep the list secret. He had just received 146 investigative reports from Europe and nearly all of them were bad. The file in front of him at that JIOA meeting contained proof that the Germans' Nazi backgrounds violated the terms of the policy that President Truman had signed. The Office of Military Government US (OMGUS) Security Report on the scientists contained histories of SS membership, war crimes, and crimes of a lesser, yet still embarrassing nature. Wernher von Braun's report had been one of the first to arrive at JIOA, with the clearly stated conclusion that he was an ardent Nazi and a security threat to the United States. The captured SS records showed that he had not only been an SS member (number 185068) since May 1st, 1940, but had also previously been a student member of an SS riding school and a Nazi Party member (number 5738692) since May 1st, 1937.

Klaus was unaware of this because JIOA held all German specialist files classified, thus preventing any independent investigation being carried out. Director Ford was supposed to give him the dossiers on 334 PAPERCLIP scientists living in the United States without visas but categorized as "enemy aliens under military custody." With a great number of those having derogatory OMGUS Security Reports, it was obvious that PAPERCLIP was a ticking time bomb.

If von Braun was the highest profile suspect, there were others at Fort Bliss who were more ardently Nazi. Director Ford had a photograph of Anton Beier, a Peenemünde engineer, fully dressed in an SS uniform, complete with the skull and crossbones on his SS cap. Beier had joined the SS in the same year that Hitler had come to power and his OMGUS file revealed his membership of the Nazi

Party and two other Nazi organizations.

Kurt Debus, a German who would eventually become the first director of the Kennedy Space Center at Cape Canaveral, was another member of the SS. In 1942 he had reported a colleague for making anti-Hitler comments during an argument over who had started World War II. The man was tried and found guilty, being sentenced to serve two years' imprisonment. (There was a slight similarity between this and the fate of Vassily Mishin's father after not reporting a joke about Joseph Stalin.) When the JIOA had asked Debus about this case at Fort Bliss, Debus had claimed that the prison term had been suspended thanks to his testimony at the trial. The OMGUS Security Report had shown otherwise. Debus' testimony had incriminated the man but the prison sentence had been suspended thanks to the intervention of the man's employer, not Debus. His claim to have only applied to join the SS was also easily dismissed by reference to the SS records at the Document Center in Berlin. They showed that he had joined the SS in 1939 and was assigned SS membership number 426559. Just as damning were the personal testimonies of his wartime colleagues who reported to the American and British interrogators that they had seen Debus wearing his black SS uniform around Peenemünde on several occasions.

Over a dozen of the rocket scientists at Fort Bliss had worked at the Mittelwerk V-2 factory at Nordhausen, including Arthur Rudolph and Magnus von Braun. Rudolph had been the production manager there and had by far the dirtiest hands when it came to the employment of slave labor from the nearby Dora camp. Rudolph was an unashamedly ardent Nazi who hated the Jews. He had been in the rocket business since the early days of Max Valier, whom he'd seen die in an explosion. He joined the Nazi Party in 1931.

"I read *Mein Kampf* and agreed with a lot of things in it," he later explained. "Hitler's first six years, until the war started, were really marvellous. They were the best years in Germany. Everybody was happy. Everybody got jobs." Rudolph marched through the streets of Berlin in his brown shirted SA uniform, singing Nazi Party songs with lines such as "When Jewish blood spurts from the knife, then all goes twice as well." At Nordhausen he managed the full-scale production of V-2s and was therefore a prime target for recruitment by the Americans under Operation OVERCAST.

With an estimated death toll of over twenty thousand slaves during the building of the subterranean factory and the subsequent production of missiles, JIOA was obviously worried about the Nordhausen connections of its rocket team in America. The No.1 German responsible for war crimes committed at Nordhausen was Georg Rickhey, the Mittelwerk general manager. He wasn't at Fort Bliss, but he was in the United States, at Wright Field. The upcoming Nuremberg trials could cause great problems if Rickhey was put on trial.

Magnus von Braun had worked as an engineer under Arthur Rudolph in the Mittelwerk from 1943 until the Germans had been forced to flee the area in 1945. His presence in Fort Bliss was entirely due to his family relationship with Wernher von Braun. If his older brother hadn't brought him to America, he would still be trying to get selected through normal channels in Germany. CIC agents at Fort Bliss rated him as "a dangerous German Nazi" because of his pro-Hitler views. "His type is a worse threat to security than a half dozen discredited SS Generals."

The arrival of the scientists' wives at Fort Bliss also brought problems for Director Ford and the JIOA. There was one wife in particular who was to cause great trouble: Ilse Axster. US intelligence reports told of her leadership of the NS-Frauenschaft—a women's Nazi Party auxiliary in charge of preaching Nazi propaganda to youth groups. Neighbors told of her wartime antics of walking through the village wearing riding breeches and brandishing a horsewhip. Witnesses told investigators that she frequently beat her Polish workers with the whip or called the police and had them beat the workers for her. Her presence in Landshut had caused those village neighbors to complain angrily to US officers in Germany that she should be arrested and prosecuted. The reason she wasn't was that she was protected by her husband's employment under project PAPERCLIP.

When the story was leaked to the press, Rabbi Stephen Wise at the American Jewish Congress took up the case. Ilse Axster was a flagrant breach of the denazification laws and her transfer to America to be with her husband proved that the so-called screening was a farce. Wise wrote a letter of complaint to the Secretary of War, Robert Patterson, who replied with a promise to have investigators look into the allegation. Rabbi Wise's letter was forwarded to the JIOA, who apparently dismissed it as nothing more than Jewish propaganda. Chairman Wev's comment that being forced to investigate the Germans' Nazi pasts was comparable to "beating a dead Nazi horse" was apt but unfortunate.

Director Ford therefore had good reason to deny Klaus the right to inspect the classified files. The newly arrived OMGUS Security Reports were now an integral part of the JIOA dossiers on each of the Germans already in America and those under consideration. It was also obvious that Klaus was not going to sign any waivers. President Truman had outlawed any entry into the United States by "ardent Nazis." The JIOA's only solution to the problem was to get the OMGUS Security Reports altered.

Director Ford sent a cable to G-2 in Frankfurt with a request for intelligence officers to verify von Braun's political background and seek out any facts that could be used as extenuating circumstances for his SS membership. Ford was looking for the new OMGUS report to state that von Braun was only a nominal Nazi Party member and that his SS rank was an honorary appointment. As OMGUS dealt with Ford's request, the whole PAPERCLIP scam was revealed by reporter Drew Pearson. In one of his regular radio broadcasts he announced that Karl Krauch, the senior IG Farben director who was currently in prison awaiting trial at Nuremberg for his part in the setting up of the chemical giant's factory at Auschwitz, had been offered a PAPERCLIP contract to work in the United States. The letter of offer had been unbelievably sent to Nuremberg prison, where it was intercepted.

Pearson's sensational exposé reached a very large audience, including the US Army Chief of Staff, General Eisenhower, who demanded an immediate explanation of how something so immoral could happen. The Chief of Military Intelligence, Lieutenant General Stephen J. Chamberlain, requested the assistance of Colonel Hagood to explain it personally to Eisenhower on March 11th, 1947. Because Hagood was about to retire, he had not been privy to Director Ford's decision to get the OMGUS reports rewritten, so his presentation to Eisenhower gave a false picture of the state of PAPERCLIP. The Army Chief of Staff was apparently convinced by the assurances that all of the Germans under PAPERCLIP were carefully screened. Eisenhower was also told of the State Department's attempts to sabotage the process through the actions and inactions of Klaus. At no time did the subject of Nazis already in America come up.

The crisis was far from over. Hagood's retirement was postponed for a further forty-eight hours in order to give him time to brief both Secretary of War Robert Patterson and his assistant, Howard Petersen. Pressure was applied to the pro-PAPERCLIP General Hilldring, head of the State Department's visa section, to finally get rid of the obstructive Klaus. All remaining visa applications would have to be rushed through since Petersen had informed Chamberlain that Operation PAPERCLIP would have to be wound up six weeks later, on June 30th, 1947. With the removal of Klaus, and his replacement by his colleague Herbert Cummings, Director Ford assumed that the process would go unhampered, but Cummings proved just as unwilling to knowingly issue visas to notorious Nazis. General Hilldring could offer Ford no assistance this time, since Cummings had spotted inconsistencies between OMGUS security reports and the actual Nazi Party records.

As matters started to deteriorate for PAPERCLIP in general, a chance discovery by a Dutch interpreter at the former concentration camp at Dachau led war crimes investigators straight to Fort Bliss. Wilhelm Aalmans was a member of a team operating out of Dachau that now served as the headquarters for the US Army's war crimes investigations and trials. He was interrogating former SS officers who had served at Nordhausen but had come up against a formidable wall of silence. With only eleven officers in custody out of the three thousand officers who had run Nordhausen and Dora, Aalmans was faced with an impossible task. Most of the inmates who had told their stories on their liberation two years previously had disappeared to the four corners of the earth.

The only clue was a telephone list salvaged from Nordhausen in April 1945, on which were two names: Rickhey and Rudolph. When the SS officers held at Dachau were interrogated and asked about these names, they denied all knowledge of them. As far as they were concerned, the Dora prisoners had killed each other. Aalmans' breakthrough came when he was reading the US military magazine *Stars and Stripes*. There, under a tiny headline of "GERMAN SCIENTIST APPLIES FOR AMERICAN CITIZENSHIP" was the name Georg Rickhey. Aalmans screamed for joy and within three days, Rickhey was back in Germany awaiting trial at Nuremberg. The former Mittelwerk general manager denied all knowledge of the killings at his factory. He was always somewhere else. Aalmans and his team were still blissfully unaware that Arthur Rudolph was also alive and well and living in America at Fort Bliss, Texas.

On May 29th, 1947, as night fell, a modified V-2 rocket took off from White Sands and headed over the Organ Mountains of New Mexico to a testing spot in the desert. It was part of Dr Richard Porter's General Electric Project HERMES, the second missile to be tested. Its trajectory suddenly changed as a result of a gyroscope malfunction and the Hermes II rocket reversed direction and headed south for El Paso. The range safety officer was Dr Steinhoff and he chose not to order the destruction of the wayward missile. The Naval Research Laboratory sailor in charge of the destruct button was told to let the rocket fly south over the city and over the Mexican border towards Juarez, a favorite haunt of the Germans on their frequent drinking trips south of the border. Steinhoff assumed that the missile would also clear Juarez, but the General Electric Hermes II ploughed into a hillside just outside the Tepeyac Cemetery, narrowly missing a building where local Juarez construction companies stored their dynamite and blasting powder. The missile did not have a warhead but still managed to punch a crater thirty feet deep and fifty feet across into the rocky hillside.

Colonel Turner anxiously checked with his friend, the commanding general of the State of Chihuahua, who reported that amazingly no-one had been killed or injured and that there had been no damage to property. The Mexican general then promised to clear the whole event with the authorities in Mexico City. No sooner had Colonel Turner put down the phone than he picked up another one and found himself explaining the whole thing to an irate General Eisenhower. Calming Ike down, he then had to deal with the Secretary of State, George C. Marshall, who in the morning would have the unenviable task of explaining to the Mexican government why the United States had decided to launch a ballistic missile on its peaceful neighbor.

Diplomatic relations were smoothed by offers of compensation from the US Army. One enterprising Mexican put in a claim that the rocket blast had made his wife sexually frigid—he eventually accepted $250 in cash.

Steinhoff was investigated thoroughly, and survived accusations of sabotage, when the board of inquiry saw merits in his avoidance of a propellant explosion over El Paso. One naval officer remarked, "I've never seen such a cold-blooded bastard." Steinhoff himself would later recall that within ten minutes of the impact, Mexican food stalls in the area were already selling still-warm souvenirs of the top-secret GE Hermes II. As with holy relics the total amount of wreckage sold was far in excess of the weight of the empty missile.

Although Aalmans and the war crimes investigation team were unaware of Arthur Rudolph's existence, the US Army Air Force visited Rudolph at Fort Bliss on June 2nd, 1947. Major Eugene Smith interviewed him in Colonel Toftoy's office and wanted to know if Rickhey had ordered the public hangings of the prisoners charged with sabotaging V-2 production. Rudolph had to box clever in order to avoid incriminating himself. It became quickly obvious that Smith knew very little about Nordhausen, so Rudolph downplayed his own role and claimed ignorance on other matters. He told Smith that up until the final months, the working conditions in the tunnels had been good, with sufficient ventilation and adequate food rations. He stressed that the production staff ate the same food as the workers. Concerning the hangings he admitted that he had gone to see the final stages of the execution out of curiosity.

Smith interviewed two other engineers then gave up when the conspiracy of silence was obvious. He never bothered questioning Magnus von Braun. Reporting back to his headquarters, Smith admitted that Rudolph was a shrewd man whose answers were structured so as to incriminate no-one. The signed statements were forwarded on to Dachau, but, forty years later, Aalmans claimed that he had never received them. If he had known that Rudolph was alive and had been interrogated, he would have summoned him to Germany to use in the trial of Rickhey. The following month, Dachau did request the presence of Wernher von Braun as a defense witness for Rickhey, but Colonel Toftoy refused to help. He suggested that they try to get General Dornberger from the British instead.

The Nordhausen trial started on August 7th, 1947, with a total of nineteen SS guards accused of brutality and murder. Arthur Rudolph's former secretary, Hannelore Bannasch, testified that her boss had passed sabotage reports to the SS and had allowed beatings of prisoners to occur. The working conditions at Nordhausen were the responsibility of Rudolph and this fact went unnoticed by the court and the press reporting it back to America. The trial lasted four months and ended in fifteen convictions and one death penalty.

On their initial search at Nordhausen, intelligence officers had found a record of a meeting dated May 6th, 1944, chaired by Georg Rickhey. A group of twenty-eight engineers, SS, and Wehrmacht officers had been summoned to discuss a problem concerning steering assemblies. There was a production bottleneck that needed to be cleared by the employment of further workers. The group decided that eighteen hundred skilled workers in France should be taken prisoner and transferred to the concentration camp at Dora. All group members agreed that the abducted French workers should be distinctively clothed in striped suits. Among those group members present were Arthur Rudolph, General Dornberger, and Wernher von Braun.

Visas

The suspicion that the Mexican crash landing was the result of sabotage was part of a general belief in a German conspiracy. Boredom at Fort Bliss had led to various breaches of security over the period, with the German scientists often heading across the Mexican border into Juarez to sample the local delights. But the FBI and Army CIC agents had greater concerns than the liaisons with Mexican prostitutes, for Mexico was swarming with communist agents working out of the embassies in the capital. Northern Mexico had been the base for Nazi spying activities during the war and German sympathizers had been interned but later released.

With Fort Bliss and White Sands Proving Ground less than forty miles from the border, the CIC counter-espionage headquarters believed that the Russians could hardly fail to be interested in penetrating the bases. Evidence for this came from, of all places, Camp Landshut where CIC agents had caught none other than Helmut Gröttrup's mother, Thea, involved in espionage activities. Together with Helmut Thiele, a well-known recruitment agent for the Soviets, Thea had been questioning returning scientists and their families about project PAPERCLIP in America.

Before the families had been allowed to reunite with their husbands, the unreasonably long delays in the delivery of mail had led Army Ordnance to bypass the system. It was more important that the morale of the specialists in Texas be maintained, especially when they were not being used for worthwhile projects. This circumvention of the standard US Army censorship allowed obvious abuses to be committed. Three Germans had somehow managed to get telephones connected up in their quarters, which were used for international, unmonitored phone calls. For those not using their own phones, the use of local pay phones was also unmonitored. This was despite the original claim by Army Ordnance that the Germans were to be the most heavily watched group in history.

A local Fort Bliss businessman reported to the FBI that one of the Germans, Hans Lindenmayr, had been using his company's address as an illegal mail drop. The FBI intercepted a letter from Lindenmayr's wife in Germany, yet she was supposed to be living with him on the base. The FBI learned that Lindenmayr had abused the PAPERCLIP system by bringing his mistress, posing as his wife, to the United States. His real wife had found out and was threatening to file a complaint with US immigration officials unless the problem was sorted out. Captain Paul Lutjens was the head of CIC branch intelligence at Fort Bliss and it was his responsibility to deal with all security breaches. He arranged to interview Lindenmayr in Major Hamill's office where the German admitted that the claims made by his wife were true but there were extenuating circumstances. Interested to hear what story the German would concoct, they were still surprised to hear his tale.

In 1938, he had been arrested by the SS and charged with espionage. He claimed that in order to avoid a one-year prison sentence, he had married the prosecuting attorney's daughter. After being forced to live with her for a few months he had deserted her for the woman who was now living with him in Fort Bliss. Major Hamill believed the story and allowed him to remain on the Rocket Team. With the general illegal alien nature of the OVERCAST/PAPERCLIP recruits, it was hardly justifiable to punish one of them who had helped a lover to enter the United States illegally. Army Ordnance decided to cut his salary by $2,000 and ordered him to file for divorce from his wife. His mistress was to return to Germany until the divorce had gone through then she would be free to re-enter the United States as his wife and dependent.

More sinister cases than that of Lindenmayr's were investigated.

Three other Germans were running illegal mail drops in El Paso where they received money from unknown foreign sources and coded messages from South America. CIC-endorsed networks existed in Argentina and other surrounding countries for Nazis who had escaped prosecution in Europe. The fascist element was frequently overlooked when their anti-communist value in the developing Cold War was assessed. Nevertheless, the CIC were unable to trace the true source of this money. One of the Germans received $3,000 in cash from Chile, and one month another deposited $953 into his bank account on top of his normal salary of $290. Even more intriguing was how more than a third of the PAPERCLIP group suddenly acquired expensive cars.

The FBI was investigating a German-born woman who ran a grocery shop in El Paso. She had been described by her neighbors as pro-Hitler during the war and the Fort Bliss Germans frequently visited her—the shop becoming a local gathering place for Americans of German descent. Captain Lutjens found that there had been so many violations of security that it was not surprising Major Hamill had never even required the Germans to fill out federal government forms applying for security clearance in the first place. Hamill still had a memo sent from the then chief of G-2 Exploitation Branch, Lieutenant Colonel Hagood, ordering the discontinuation of the surveillance four months after the Germans had entered the United States.

It was no wonder that people suspected the Germans of sabotaging the V-2 launches at White Sands. The rocket that had narrowly missed hitting Juarez was not the only one to go astray. Within a period of one month, three out of four missiles fired from White Sands had overshot the ninety-mile-wide range and landed off-course near heavily populated areas in New Mexico. Apart from the Juarez incident, one landed four and a half miles south of Alamogordo and another crashed only two miles from Las Cruces. Cameras on board one of the missiles were never recovered—neither were parts from another. When Lutjens investigated these incidents he found that the army and navy criticized each other over the shared rocket range, but did not blame the Germans. Hamill himself blamed Dr Richard Porter's Project HERMES team from General Electric.

General Electric replied with a criticism of the Army Ordnance lax security at White Sands, which the GE manager, A.K.Bushman, described to FBI Director J. Edgar Hoover as bordering on "criminal neglect." He was particularly concerned about the former Nazis having access to General Electric's own classified information such as the new rocket fuel being developed at its missile project in Malta, New York. Bushman informed Hoover that he didn't trust the Fort Bliss Germans and their contacts with others back in Germany. It should not be forgotten that, according to GE information, at least 350 of their former colleagues were now working for the Russians and the possibility of communication between the two groups should not be dismissed as easily as Army Ordnance lack of surveillance implied. The rift between General Electric and the Germans at Fort Bliss would have severe repercussions during the next decade as the race to develop a satellite began.

Suspicions also were levelled at Wernher von Braun when it was discovered that he had sent a map overseas to his family. American officers in Germany were still trying to find more V-2 documents. Although the larger caches had already been sent to the Foreign Document Evaluation Center at the Aberdeen Proving Ground in Maryland, von Braun's map was to indicate the hiding place of a cigarette box stuffed with sketches. This small treasure cast doubt on von Braun and his willingness to be open with his American employers—especially Major Hamill, who had risked his own life just after the war by leading a snatch team into Russian-occupied Poland to rescue von Braun's parents.

The map was to be given to the wife of General Dornberger, who would search for the documents since her husband was still in a British prison. She was to hand them over to a German scientist who was on the list of next PAPERCLIP recruits to travel to the United States. Once in von Braun's possession, they would be used as a bargaining chip for getting Dornberger into the program. However, the scheme failed when US officers found and confiscated the map from Dornberger's wife. The Americans even managed to arrange Dornberger's release from the British on July 9th, 1947, and fly him to Germany to search for the treasure. Dornberger had them wandering around a forest for days, digging up tree stumps but finding nothing. Eventually, when they were found, they were nothing more than rotten pieces of paper, destroyed by rain during the two years since their burial.

With Dornberger's release from prison, the Peenemünde team could have been reunited in Texas, but it was still felt that Dornberger posed a threat. Dr Howard Robertson's view still stood:

> He wields great power over his subordinates, including Prof Von Braun. I am convinced that Dornberger is a most dangerous man, and that he should in any case be shorn of all influence over and even prevented to have contact with his former Peenemünde subordinates.

Having a former Nazi general in charge of civilians at Fort Bliss would have been a public relations nightmare—one which the Army Ordnance did not want. Dornberger did enter the United States under PAPERCLIP but was employed by the newly created US Air Force as a guided missile consultant at Wright Field, Ohio.

The British were glad to be rid of General Dornberger. They had other, more manageable Germans working for their missile program. The Operation BACKFIRE at Cuxhaven had inadvertently aided the Americans, not through any information it received but by its temporary employment of Arthur Rudolph. The Canadian Field Service (CFS) had uncovered evidence that Rudolph was the production director at the Nordhausen Mittelwerk and had been responsible for the foreign slave labor there. When the CFS officers demanded to interrogate Rudolph at Cuxhaven, their request was denied for reasons of security. Thus, Rudolph was spared a very early incarceration as a possible war criminal.

The British missile program was hampered right from the start by the views of its pre-eminent homegrown rocket expert, Sir Alwyn Crow. It must be remembered that back during the war, when Professor Lindemann was fiercely denying the existence of the V-2, he was supported by Crow, whose only experience had been in solid-fuel rockets. Crow had developed basic but effective short-range two- to five-inch rockets powered by solid cordite fuel, and if the Nazis were using the same fuel it would require a missile weighing seventy tons to lift a one-ton warhead a distance of over a hundred miles.

Crow's greatest fault was his persistent unwillingness to consider the possibility that the Germans were using liquid fuel. As early as August 1942, he had visited Shell's laboratories in Horsham where the engineer Isaac Lubbock was experimenting with liquid fuel rockets. Despite witnessing a successful experiment, Crow ruled out the possibility of discovering an effective way to pump liquid fuel into the combustion chamber fast enough. Over in Peenemünde, Walter Thiel had already solved the problem with a turbo-pump driven by a gas turbine. (Von Braun's expert designer was the only major engineer to be killed during the RAF raid on Peenemünde the next year.)

Once the V-2s had proved their existence to Lindemann and Crow by raining down on London, Crow's deputy, Colonel William Cook, and another solid-fuel adherent, John Elstub, were sent over to Europe as CIOS Team 163. Despite their attempts to obtain V-2 parts at Nordhausen—frustrated by the antics of Staver, Hochmuth, and Edwin Hull—and Cook's interrogations of von Braun and the team at Oberammergau, neither man was invited to attend the launchings of the Operation BACKFIRE V-2s at Cuxhaven. Two other key personnel in the British missile program were also conspicuous by their absence. One was Ben Lockspeiser, a chemist who was the chief scientist at the Ministry of Aircraft Production during the war; the other was Sir Alwyn Crow himself. It begins to look as though the British missile men were just as uninterested in the Cuxhaven demonstrations as the Americans were.

With his half-hearted attempt to recruit von Braun during his visit to London, Crow seemed to give up any chance of building a decent British missile research organization. Cook and Elstub had chosen an abandoned airfield at Westcott, northwest of London, as the research center, but were denied access to British personnel files for recruiting purposes.

William Cook summed up the sorry state of British attempts to acquire German specialists: "Right at the beginning, we were already too late. The best rocket scientists had already gone to America and Russia. They had been so much faster off the mark. We sent search parties to find those who had not yet been hired, but it proved impossible...from the outset, Crow and Lockspeiser dampened our enthusiasm because they just did not understand the range of developments which the Germans had unlocked."

The best that the British could round up were a dozen BACK-FIRE recruits who were accommodated in a hotel on Luneberg Heath. They continued their research on rocket fuels at the underground rocket test base at Traun. "Papa" Riedel was a Peenemünde veteran whose experience in rocketry went all the way back to the days of Max Valier. In fact, it was Riedel who had held the dying Valier in his arms after the explosion. Despite his historic status, the Americans had refused to employ him, so the British picked him up. But the real leader of the German group was a newcomer to ballistic missiles: Dr Johannes Schmidt. His previous work had been on the rocket booster for the Messerschmitt Me163 interceptor. As far as V-2s were concerned, he had never worked at Peenemünde.

With the replacement of Sir Alwyn Crow by Ben Lockspeiser, the situation appeared to get worse. Cook's opinion of Lockspeiser was not good: "He was an aircraft man who didn't understand the difference between planes and rockets, and wrongly thought that the design work was just the same." With the closure of the Traun test base in mid 1946, under Allied Law No.25 ordering the dismantling and destruction of all German wartime research, the Germans were either to be abandoned or relocated to Britain. Eventually a decision was made and the twelve Germans arrived in Britain on, of all nights, Bonfire Night, November 5th, 1946. With the skies of Britain full of fireworks in remembrance of Guy Fawkes' seventeenth-century attempt to blow up the Houses of Parliament in Westminster, the Germans were greeted with what could have been comically described as the best that British rocketry had to offer.

At Westcott, the Germans were not billeted in a comfortable hotel as they had been in Luneberg Heath, but in old Nissen huts in a barbed-wired compound on the airfield. There was also understandable hostility from the local English population, and the barbed wire was as much for their protection as the barbed wire at Gorodomlya Island in Russia had been for the Germans there. For the next year, the team worked on John Elstub's liquid oxygen anti-aircraft missile, which was fatally flawed. The project was plagued by faulty components, engine failures, and fuel problems. The final blow came on an unlucky Friday 13th. During a test firing on November 13th, 1947, the rocket exploded, killing the team leader Dr Johannes Schmidt. With his tragic demise, the British rocket program ground to a halt. Both Cook and Elstub had had enough of Lockspeiser and the constant Whitehall interference, and duly resigned.

In the great race to grab German scientists, Britain had never stood a chance. Even with America and the Soviet Union rushing ahead of them, they failed to hold on to third place. The fourth Allied Power, France, had been far more successful in recruiting rocket specialists. During the war, Peenemünde had used several French companies to provide components for the V-2, and after the liberation of France, Allied CIOS teams were quick to search for suppliers.

Understandably, those companies which had collaborated with the Nazis wanted their involvement in rocket technology to be kept secret. CIOS teams were frequently refused access to the classified documents. The matter of non-compliance was raised by General Eisenhower in a letter to the French government. General Alphonse Juin, chief of the General Staff, replied that the CIOS teams should refrain from "asking any questions or conducting any inquiries which might appear to be aimed at obtaining information of a confidential character on French industries." With such a blatant snub, the French General was still cheeky enough to ask that French Scientific Coordination Committee teams be allowed to join CIOS teams in Germany. This was refused since "nothing is to be gained by French participation."

British CIOS members decided that whatever V-2 secrets could be found in Germany should be denied to the French, treating them with as much distrust as the Russians. Brigadier-General Eugene Harrison, a G-2 officer in the US 6th Army Group, discovered evidence confirming British and American fears about France. While accompanying the French Army in southwest Germany, Harrison had learned that French intelligence agents, working under the cover of the Sécurité Militaire, were ordering and sometimes threatening German scientists not to cooperate with CIOS teams in the area. The threats usually took the form of charging them with war crimes if they cooperated with the Americans and British.

Harrison came across cases where the wives of scientists were arrested and forced to blackmail their husbands into returning to the French zone of occupation. As the extent of the French success in acquiring scientists and documents became known, America and Britain attempted to take a softer line by agreeing to share CIOS intelligence in the hope that some French assets would also be shared. Complete trust was never established since the names and data on the CIOS Black List was not passed on to General Juin.

A large group of German rocket scientists were contracted to reassemble V-2s at St Louis in the Alsace region of France, and allowed to return to their families and homes in Germany each night. Lieutenant-Colonel Henri Moureau had secretly recruited many of these while he was working for the British at Cuxhaven, during Operation BACKFIRE. St Louis was perfect for the French effort at reassembling the V-2s since the Germans could live across the Rhine in the two villages of Weil and Haltingen and be bussed into France each day. The operation was so efficiently run that the potentially angry local population never knew what was going on. The proximity of the three university scientific libraries of Strasbourg, Basel, and Freiberg was also an advantage.

In the summer of 1946, three hundred of the German scientists were relocated to Vernon, Normandy—the site of France's new rocket center. Wolfgang Pilz, a lesser luminary at Peenemünde, now became a key player in the designing of the French Army's *Veronique* rocket. His thoughts were also of space travel and the future plans of the European *Ariane* satellite launch vehicle began at Vernon. The famous couple Eugene Sänger and Irene Bredt—designers of the "antipodal bomber" that was supposed to be able to reach New York and so interested Stalin that he ordered Soviet intelligence agents to kidnap the pair—were married in Paris and spent their honeymoon at the French Air Force Aeronautical Arsenal at Chatillon.

Of course, not all of the so-called German rocket scientists were actually German. One interesting case was that of Hermann Oberth, one of the great triumvirate of Space Rocketry that included the American Robert Goddard and the Russian Tsiolkovsky. He had been born in Transylvania in 1894 and published the seminal *The Rocket into Interplanetary Space* in 1923. Forced to take German nationality to avoid the concentration camp, he had worked briefly at Peenemünde, but because of his Rumanian birth was denied access to the inner workings of the V-2 development team under von Braun. Ironically, he was given the chance to develop a solid-fuel rocket at a time when everything was liquid-fuel.

Leaving Peenemünde for Wittenberg on the river Elbe, his project was soon frustrated by Allied bombing. After a period of postwar internment, and rejection by the American PAPERCLIP recruiters, he found work with the Italian Admiralty at their research establishment at La Spezia. It was there that he finally developed his solid-fuel rocket. For such a living legend to end his days in the backwoods of rocket research was itself a crime. The Americans had wasted the talents of Robert Goddard. Von Braun was not going to let Oberth suffer a similar fate.

The Germans at Fort Bliss had been joined by their wives, who were able to enter the United States legally. This was still not the case with the German scientists themselves, for their visas were still being withheld by the State Department. The whole problem should have been resolved by the end of the financial year—June 30th, 1947—but PAPERCLIP continued to exist beyond the date given for its winding up. The PAPERCLIP office in Germany operated out of the Intelligence Division of the European Command (EUCOM) headquarters in Heidelberg. It was run by Deputy Director Colonel Robert Schow who, with the aid of his assistants Colonel William Fagg and Colonel C.F. Fritzsche, signed many of the OMGUS Security Report forms that were causing JIOA such headaches.

JIOA Governing Committee chairman Bousquet Wev sent a memo to the Director of Intelligence at EUCOM in Berlin, Robert Walsh, asking that the OMGUS reports on fourteen individuals be reviewed and "that new security reports be submitted where such action is deemed appropriate." Wev stressed that there was little chance of any of the Germans on the list being approved by the Departments of State and Justice if OMGUS classified them as even a potential security threat. If immigration was denied, then these rocket specialists would be returned to Germany and actions would have to be taken to deny their knowledge to other nations for security reasons. Wev also pointed out that the OMGUS reports were "unrealistic" since none of those listed had been politically active. Among those listed were Herbert Axster, accused by OMGUS of starving foreign workers; Anton Beier, whose SS file apparently was over two inches thick; ardent Nazi rocket technicians Guenther Haukohl; Hans Friedrich; and Wernher von Braun.

Walsh handed the request on to Colonel Schow in Heidelberg and the OMGUS security reports were duly altered. All the "ardent Nazi" categories were changed to "not an ardent Nazi." Most of the reports were signed on the same day, by Colonel Fagg and other intelligence officers. Wernher von Braun's original September 18th, 1947, OMGUS security report noted that the "subject is regarded as a potential security threat to the Military Governor." Five months later, his new report stated that since he had been in the United States for more than two years, if his conduct had been exemplary

"he may not constitute a security threat to the US." The embarrassing incident over the Dornberger treasure map was not added
to his dossier.

Axster's first OMGUS Security Report had been damning: "he
should—ideologically speaking—be considered a potential security
threat to the United States." Six months later his report was radically changed to: "Subject was not a war criminal and was not an
ardent Nazi. The record of Herbert Axster as an individual is reasonably clear and, as such, it is believed that he constitutes no more
of a security threat than do other Germans who have come to the
US with clear records in entirety." This doctored report held despite
further evidence emerging from war crimes investigations in Europe.
Even more amazing was the way that OMGUS now dismissed Anton
Beier's twelve-year membership of the SS and his photograph wearing a Death Head Division cap as mere opportunism.

The State Department's director of Office Controls, Hamilton
Robinson, was responsible for the Visa Division, counter-intelligence, and security. He was also put in charge of the PAPERCLIP
problem and, as such, became the target of JIOA. Chairman Wev fed
information to his right-wing ally in Congress—Congressman Fred
Busby from Illinois. On March 10th, 1948, Busby called Robinson
before a House subcommittee and charged him with incompetence
and having family links with a communist. Busby went over the top
in his allegations and the other members of the subcommittee had
to shut down the hearings, but this only freed him up to continue
the attack on Robinson in news broadcasts over NBC radio.

NBC reporter Ned Brooks quoted Busby as his source when he
accused Samuel Klaus as being "the man most influential in sabotaging the program" to give German scientists visas, despite
President Truman's approval of PAPERCLIP. Busby took the fight
against Robinson to Congress with a long speech accusing the State
Department of allowing its visa section to be run by communists.
Busby succeeded, and within two days, Robinson had offered his
resignation. The road was now clear for JIOA to process the immigration of former Nazis.

On May 11th, 1948, General Chamberlain, the director of intelligence, made a presentation to FBI Director J. Edgar Hoover on the value of the German scientists to the national security of the United States. Raising the fear of communist acquisition of technological talent unless it was denied them, Chamberlain succeeded in convincing Hoover to overturn his recommendation of September 13th, the previous year, that put an end to German access to classified technical information. As a result, Hoover promised to meet personally with the attorney general to expedite the visas.

The process by which the visas were eventually granted was itself an interesting one, for the Germans had entered the United States illegally almost three years before. As far as the State Department was concerned, they were not even registered as being in the country. To rectify the situation, the Germans were herded onto rail cars in El Paso and taken across the Mexican border to Juarez, the site of their previous missile "attack." The American consulate there issued them with visas to enter the United States legally. Their immigration papers bore the strange statement that the port of embarkation was Ciudad Juarez, Mexico, and the port of arrival was El Paso, Texas—neither city is within seven hundred miles of the sea.

As the Germans acquired legal status in Texas, major events were occurring in the Soviet Union. The Russians had decided to move Gröttrup and his team from NII-88 to Gorodomlya Island, where the other German group were based. Gröttrup was not particularly sad at leaving the dilapidated factory at Kaliningrad but the fact that his salary was cut down to five thousand rubles was a sign of his decreasing importance. Gorodomlya facilities consisted of six wooden two-floor apartment buildings with an inadequate water supply and sewerage system. The German rocket team occupied the bottom floors and the Russian personnel the upper floors. Not only was the island surrounded by barbed wire, but armed guards were posted at the two piers on the northern and southern ends of it. Every fortnight, German housewives were allowed an escorted shopping trip to the town of Ostashkov, on the southern shore of Lake Seliger, where market prices shot up accordingly.

Despite the standard of the accommodation, the German wives had made efforts to transform the settlement into a little piece of Germany. The men were provided with sophisticated development and testing facilities—including a Mach 5 wind tunnel, electronics lab, and a propulsion test stand—yet their G-1 project ground to a halt as the Russians passed judgment on their designs. They were not as innovative as the Korolev team had hoped. Gröttrup learned that Korolev had been given the go-ahead to abandon the R-1 and develop the Russian improvement on the V-2, designated the R-2. The Russian innovations would not be as great as the G-1.

On April 9th, 1949, Minister of Defense Armaments Ustinov paid a visit to Gorodomlya Island and set them a new challenge: to replace the G-1. Ustinov wanted them to design a rocket capable of sending a three-ton warhead over a distance of two thousand miles. Unknown to Gröttrup, this was exactly the same challenge that had been set Korolev for the next Russian rocket, the R-3. The Germans were pessimistic about increasing the range of a V-2 by just stretching its length; a radical rethink of the design of a rocket's shape was required.

Having considered multi-stage variations, they decided to concentrate on a single-stage cone-shaped rocket. Breaking from tradition was possible, Gröttrup reasoned, because his team had not gone through the Peenemünde school of rocketry. He speculated that Wernher von Braun would have difficulty exploring new avenues in America. He now viewed his Gorodomlya team of Wolff, Magnus, Albring, and Umpfenbach as more capable in the future field of intercontinental ballistic missiles than von Braun's.

The G-4 would require an engine thrust of a hundred tons and the Germans were given three months to come up with the design. Gröttrup was optimistic that the project was a sign that the Russians were finally going to involve them in real research and development, but this was soon shattered when the German papers were collected and no-one was invited to present the designs in Kaliningrad. Another team of officials arrived at Gorodomlya in

August 1948 to request more details of the G-4. This process of requesting plans without offering feedback continued over the next couple of years. At one point, in October 1949, Minister Ustinov reappeared with Korolev to discuss certain points.

This was the first time that Korolev had journeyed to Gorodomlya, and Gröttrup correctly interpreted this appearance of the Chief Designer as indicating that NII-88 were now extremely interested in intercontinental rockets. More than that Gröttrup did not know, for the Soviets had very definitely chosen to isolate the Germans and deny them exposure to current Soviet research. Gröttrup's team soon became despondent as they realized that they were never going to physically work on the project or see their creation fly. The close-knit community at Gorodomlya put extra pressure on marital relationships as more and more cases of infidelity and adultery appeared. The talented, frustrated men turned to the readily available vodka, and one engineer named Möller wandered out into the snow to commit suicide, but only succeeded in losing both hands to frostbite. Engelhardt Rebitzki's wife was more successful: she hanged herself indoors.

In April 1950, Minister Ustinov made a decision to phase out the missile development work at Gorodomlya and on August 13th, the Soviet of Ministers issued Decree No. 3456 which effectively established conditions for the repatriation of the Germans. This return of the scientists to Germany was top secret but the process would be a lengthy one. The Soviets adopted the policy of isolating the Germans from new Russian technological developments so that when they were eventually released, any Western intelligence to be gained from their debriefing would be several years out of date and therefore useless. While denying them access to current information, the Russians still used them as "walking dictionaries" according to Gröttrup. The first sign of this was the sudden construction of new apartment buildings on the island and the appearance of young Russian engineers who teamed up with the Germans to learn all they could as part of a general education in rocketry.

On December 2nd, 1950, the NII-88 began research into new

types of propellants that could be stored on board rockets for prolonged periods. This project, designated N2, was requested by the Soviet military in order to find a replacement for liquid oxygen that quickly evaporated from the propellant tanks. Another propellant was needed which would increase the operational readiness of a missile and the N2 search was sent to Gorodomlya as well as NII-88. The suggested replacement for liquid oxygen was nitric acid. When Gröttrup learned that the Russians were expecting him and his team to test fire the highly toxic nitric acid, he refused. On December 21st, Gröttrup was dismissed as chief of the German collective on Gorodomlya.

When all hope of ever returning to their homeland seemed to have gone and the Germans were resigning themselves to their new role of educators to the ever-increasing flow of Russian engineers, the Soviet authorities suddenly announced that a group of twenty German technicians would be repatriated on March 21st, 1951. They were the least qualified of the Germans, but this move brought hope to the rest. The Soviets did not decide on the future of the rest until later that year, in September. At the same time, Gröttrup's status took another dive. His salary dropped further, and he was asked to vacate the four-bedroom house with the only indoor bath on the island and move into a smaller apartment. Gröttrup had started drinking to deal with the humiliation and in February of 1952, he fell seriously ill. For a fortnight he was comatose with a very high temperature. When a doctor was eventually found, Gröttrup responded to drugs and slowly recovered some of his health.

A second group was told in June 1952 that they would be leaving for Germany. Only the top twenty rocket specialists remained, and Gröttrup could draw comfort from not having slipped out of that category. But his enthusiasm was gone. When news reached Gorodomlya on March 5th, 1953, that Stalin was dead, the Russians greeted their German colleagues with "Stalin kaput, good for you."

Maybe, at last, Gröttrup and his family would return to their homeland. There were vague promises that the issue of repatriation would be dealt with by 1954, but with the fresh influx of

German aeronautical scientists from Junkers and Heinkel to the island, the signals were confused. Finally, an official commission from Moscow arrived in November. A statement was read out: "All German specialists with the exception of twelve are to return to their country on November 22nd, 1953. They must leave within two days of that date. We take this opportunity to express our thanks for the work done."

Gröttrup was not one of the final dozen. Together with his wife, Irmgard, and children, Ulli and Peter, Gröttrup crossed the river Oder into Frankfurt-an-der-Oder on November 28th, 1953, seven years after his deportation. Some of the Germans who had preceded him over the last couple of years had carried on their journey into the West and into the arms of the US and British intelligence services. Under the operation code-named DRAGON RETURN, these scientists were extensively debriefed, as, in fact, the Soviets had anticipated.

The following are excerpts from Gröttrup's declassified file from the British Directorate of Scientific Intelligence and Joint Technical Intelligence Committee, dated December 30th, 1953:

> Gröttrup is the most important German rocket expert interrogated so far under Operation DRAGON RETURN. He was a fanatical Marxist, and his stated reason for giving information to the Western Powers about his work in Russia is his belief that Russia is an Imperialistic State and therefore an enemy of Democracy. The Information Research Department (IRD) will be interested in this aspect of the problem. Gröttrup's wife has given some information about potential defectors that has been passed to SIS.

Gröttrup and his wife arrived in London on January 14th, 1954, after having been properly documented as West German citizens. The main reason for their travel to Britain was to have talks with rocket experts from the Royal Aircraft Establishment at Farnborough and for the IRD to use them for propaganda purposes if it so desired. Other experts in the field who met with Gröttrup were Duncan Sandys, veteran of the V-2 attacks on London, and Lt Colonel Tokaev, the man charged by Stalin with the kidnapping of German scientists in the American zone and now a Soviet defector. Professor Norman, from King's College London, wrote a report dated January 25th, on his impressions of the Gröttrups:

> He did not know very much about Russia or the Russians. He was very much cut-off, met mostly Germans and he knows surprisingly little Russian. He admitted that he could not read the language with any ease. The Russians who "contacted" him obviously collected anything he produced, and that was the last he ever heard of it. He had hardly any general knowledge of Khazakstan. He seemed to me to have been sent straight to the firing ground and straight back again. On general consideration I should have thought it was extremely unlikely that the Russians would have let him go, had he known anything. After all, they are well aware of the fact that we are watching for returnees! I do not think he is a plant. He was obviously most impressed that he was received by Mr Duncan Sandys and very much impressed by the Minister's fluent and idiomatic German. His wife is intelligent, older than he is, I should say, obviously several cuts above him socially, and she appeared to be rather more anti-Russian than he was. He was not pronounceably anti-Russian himself. He was obviously embarrassed by being pressed for probable reasons that led to his release by the Russians, and I formed the impression that he was rather piqued and that his self-esteem had been wounded. On the whole I think we have got all we are likely to get and I should imagine we ought to disembarrass ourselves at the earliest opportunity.

Professor Norman's recommendation was acted upon and Helmut Gröttrup was returned to Germany. His wife, Irmgard, would eventually write a book about her experiences in the Soviet Union, entitled *Rocket Wife,* and this was published in the year following the launch of *Sputnik.*

When the first Russian intercontinental ballistic missile was launched a few months before *Sputnik*, and Western intelligence confirmed the reports made in *Pravda*, one high-ranking official at NATO headquarters exclaimed, "We captured the wrong Germans!" With the *Sputnik* launch itself, and the immediate joke from Bob Hope about "their Germans" being better than "our Germans," it can be seen how the presence of ex-Nazi rocket scientists within the Soviet Union and within the Soviet space program was a popular misconception. Gröttrup was one of the last Germans to leave Russia. The dozen that remained were electronics experts, not rocket scientists. Although the Germans were out of the Soviet Union by 1954, they had in effect been isolated from the Korolev's research and development program since Gröttrup was moved away from NII-88 and sent to Gorodomlya Island back in early 1948.

In recent histories of the Russian space program, especially German ones, there have been attempts to raise the value of the German influence on the development not of the rockets, but the engines. The work was based at Glushko's design bureau OKB-456 at Khimki, where the Germans were present for a period of four years. During that time, it is claimed, Glushko's water-cooled KS-50 combustion chamber, affectionately known as the "Lilliput," was really based on the blueprints of Werner Baum's team. From this, the larger ED-140 combustion engine that utilized liquid oxygen and kerosene, as opposed to the German mixture of liquid oxygen and alcohol, was developed. This departure from the original V-2 propellant fuels was a major development in the independence of Soviet technology. Even accepting the German influence on the fiercely anti-German Glushko, it can still be seen that the Soviet space program was totally independent of German advisors many years before the launch of *Sputnik*.

PART III: SATELLITES AND THE SPACE RACE

ORBITER

America's civilian space program can be traced back to a private meeting held in the house of physicist James Van Allen in Silver Spring, Maryland on April 5th, 1950. The gathering of a small group of scientists was to honor the eminent British geophysicist Sydney Chapman, who was visiting the Washington area. Also present were: Lloyd V. Berkner, head of the Brookhaven National Laboratory on Long Island; S. Fred Singer of the University of Maryland; Ernest H. Vestine of the Department of Terrestrial Magnetism of the Carnegie Institution; and J. Wallace Joyce, a navy geophysicist and State Department advisor.

After dinner the men started to discuss, among other things, the question of how to obtain measurements of the earth and the upper atmosphere from high altitudes. Berkner suggested that maybe it was time to stage another International Polar Year. The last such international scientific exercise had taken place in 1932, fifty years on from the first International Polar Year in 1882. Berkner was suggesting the shortening of the interval between these major events—when the scientists of many nations pool their resources. Instead of just studying polar conditions, he suggested that the third should also study the upper atmosphere. The fifty-year interval would be reduced to twenty-five years to coincide with a major predicted astronomical event: the period 1957-58 would be one of maximum solar activity. Scientists from around the world would study the atmosphere in a way which would show that the scientific world was united and not subject to the fragmentation of the Cold War.

The tiny group of scientists at Van Allen's house agreed that the next International Polar Year should be linked with a major scientific advance in order to grab the attention of the world's press. The high altitude studies would require the launching of the world's first artificial satellite. What better symbol of world peace than the peaceful conversion of ballistic missile technology? In a totally unscientific move, the period of global observation would eventually be extended to cover eighteen months and the "year" was renamed the International Geophysical Year. The idea that was hatched at Van Allen's home would become the cooperative effort of sixty-seven nations under the control of the International Council of Scientific Unions.

On October 4th, 1951, exactly six years to the day before the launching of *Sputnik*, the *New York Times* reported that a Soviet rocket engineer was claiming the Soviet Union had missile technology at least equal to that of the United States and that his nation could very well be launching satellites in the not too distant future. The Soviet engineer quoted was Mikhail Tikhonravov. It took a long time before the Americans responded, but when they did, it was in the form of a series of articles published in the popular magazine *Collier's*. The contributors were: Joseph Kaplan, professor of physics at UCLA; Fred Whipple, chairman of the astronomy department at Harvard; Heinz Haber of the US Air Force Department of Space Medicine; writers Willy Ley and Cornelius Ryan; and last, but not least, Wernher von Braun.

"MAN WILL CONQUER SPACE SOON" was read by as many as 15 million people and the articles would become the inspiration for three popular television films produced by the Walt Disney studios: *Man in Space* (1955), *Man and the Moon* (1955), and *Mars and Beyond* (1957). Von Braun would act as technical director on all three of these, but his prognostications in *Collier's* magazine were more down to earth. He described the potential establishment of a manned space station with its occupants spying on the world below. "Nothing will go unobserved," he claimed. Photographs taken from space would be as revealing as the best of the then current aerial reconnaissance photos. "Troop maneuvers, planes being readied on the flight deck of an aircraft carrier, or bombers forming in groups over an airfield will be clearly discernible. Because of the telescopic eyes and cameras of the space station, it will be almost impossible for any nation to hide warlike preparations for any length of time."

Unknown to von Braun, the US Air Force were already working on the requirements of a military spy satellite. The photographs had to be of sufficient quality and magnification that potential military targets could be spotted. The ultimate aim was to provide continuous daytime coverage of the Soviet Union. The classified RAND study, "Utility of a Satellite Vehicle for Reconnaissance" failed to convince the US Air Force that the quality of the photo coverage was good enough to warrant the investment in developing the system. Von Braun's image of space spies with their powerful telescopes pointing down to the earth's surface owed more to the science fiction of the 1950s. It would take years before satellite photography would replace good old-fashioned aerial reconnaissance.

Mikhail Tikhonravov's comments about the advanced state of Soviet space technology were taken seriously by those who had learned about the Russian programs through the returning Germans. Operation DRAGON RETURN, run by the British Air Ministry and the US Air Force's Air Technical Intelligence Center, had gleaned reports of the missile base at Kapustin Yar. An important conference of military, scientific, and industrial personnel was held at Wright-Patterson Air Force Base in Dayton, Ohio, in August 1952, to discuss the findings. The local consultant at the base was General Dornberger who had also leant a hand with the debriefings of returning German engineers and technicians. He warned those present that the Soviet missile program appeared to be in the lead. That could only mean that their space program was also ahead of the American one.

Earlier that year, the American Rocket Society had reviewed suggestions made by its members about how the prospects for space flight could be improved. This had been prompted by an apparent lack of official interest from the US government. One of the society's members answered that the best hope for an effective American space program might lie in the Soviets launching a satellite into space first. "The greatest utility with respect to civilization might be if the Russians were to build it instead of ourselves." The significance of this pessimistic view was made even greater by the fact that the member who wrote it was Dr Richard Porter.

President Truman, far from being disinterested, asked the Manhattan Project scientist Aristid V. Grosse to investigate the possibility of building a satellite. Of the people and places he visited, one was Wernher von Braun and the new home of the Rocket Team in Huntsville, Alabama. The Germans had left Fort Bliss in April 1950 and been transferred en masse to the US Army Ordnance Department's Redstone Arsenal there. The Germans welcomed the move from the deserts of Texas and New Mexico to the "Watercress Capital of the World." Fed up with living in military barracks in Landshut and Fort Bliss, the wives were especially determined to make Huntsville their home rather than just a place to live.

Denied home loans by the local First National Bank, the Germans took matters into their own hands and pooled their financial resources to purchase thirty-seven acres on the crest of Monte Sano Mountain. The German enclave would eventually house fifty family plots. Others, including von Braun, chose to develop lots in the valley, nearer the Redstone Arsenal. Huntsville was not a place to launch rockets, but they could be moved on large barges down the Tennessee River to the Atlantic Ocean and then along the coast to the Joint Long Range Proving Ground at Cape Canaveral on the Banana River in Florida.

US Army Ordnance had decided that von Braun's team was to develop ground-to-ground ballistic missiles for carrying the nuclear warheads which were the result of the Manhattan Project. It was therefore a symbolic meeting of minds when Professor Grosse visited von Braun to ask him about the possibility of launching a satellite. The results of his presidentially requested investigation were not available until August 1953, long after the replacement of President Truman by Dwight D. Eisenhower.

The "Report on the Present Status of the Satellite Problem" warned that since the Soviet Union had trailed the United States in the development of both the atomic and hydrogen bombs, it might make a serious effort to take the lead in the development of a satellite. Being first into space "would have the enormous advantage of influencing the minds of millions of people the world over." He also declared: "If the Soviet Union should accomplish this ahead of us, it would be a serious blow to the technical and engineering prestige of America the world over. It would be used by the Soviet propaganda for all its worth." Unaware of the RAND study reports, Grosse also concluded that a satellite with television cameras would be "a valuable observation post."

Until then, ordinary photoreconnaissance would have to do if the Western intelligence agencies were to gain any idea of what Kapustin Yar looked like. A joint British and American intelligence report dated July 1953 informed policymakers that the Russians, building on the V-2 technology, were in the process of developing a medium-range missile that could reach any target site in Europe. The report predicted that by 1956, Moscow was likely to have a long-range missile that could reach London, and that only a few years after that Washington would be reachable.

It was imperative that the West photograph Kapustin Yar. British intelligence agents posing as archaeologists in nearby Turkey had first detected the missile site. To get a closer look, the British decided to modify an RAF B-2 Canberra bomber for a daring daylight spy flight into the Ukraine. The Canberra was a great advancement on the planes used to photograph the test-base at Peenemünde. It could fly at a ceiling of 48,000 feet at a maximum speed of 540 miles per hour. With a range of about 3,600 miles, the plane would have to have this increased because there would not be opportunity for refuelling.

So, with extra fuel tanks installed in its bomb bay, the Canberra bomber took off from Giebelstadt in West Germany in late August 1953. Flying in daylight was necessary because of the reconnaissance cameras on board, but this would expose the pilot and plane to enemy attack. Despite being hit by gunfire from pursuing Soviet fighters, the Canberra managed to take photographs of the secret missile base and fly on to Iran where it landed safely. The mission, however, was not a success, because the reconnaissance cameras had taken blurred photographs due to the vibrations caused by damage inflicted during the Soviet pursuit.

The desire for a photoreconnaissance capability not dependent on vulnerable overflights of the Soviet Union but relying on a satellite in a safe orbit was the driving force behind a new RAND study called Project FEED BACK. At an estimated cost from inception to launch of $165 million and a projected time span of seven years, the RAND study acknowledged that several technological barriers were going to have to be overcome. Not only were the photographic quality of the cameras, both still and television, to be advanced way beyond current levels, but the electronic transmission of the pictures to earth would require new developments in data processing. And that was just the satellite—a new rocket booster would also be required.

RAND also raised the legal problems that such a project would cause, since the passage of an American satellite over enemy territory would bring charges of violation of Soviet sovereignty. The report warned: "Probably the Soviets would do everything in their power to make the charge stick and to exploit it to their political advantage by making the United States appear as a violator of international law." This last point would become a very significant one in the later stages of the actual development of the US satellite program, and would figure as the cornerstone of one particular conspiracy theory about the US failure to beat *Sputnik*.

After his early appearance in the *New York Times*, Mikhail Tikhonravov found little exposure for his ideas on space travel within the Soviet Union. While working at NII-4, a military institute which speculated on the application of ballistic missiles, Tikhonravov had authored several reports on space launch vehicles and satellites, but no action had been taken to develop them further. Chief Designer Korolev was an avid reader of his friend's R&D reports but the opportunity to help did not arise until May 20th, 1954, when Korolev was formally ordered by the Soviet government to develop the first intercontinental ballistic missile (ICBM), designated the R-7. He wasted no time in sending off a copy of Tikhonravov's latest technical treatise entitled "Report on an Artificial Satellite of the Earth" to the Soviet government with an attached letter saying:

> I draw your attention to the memorandum of Comrade M.K. Tikhonravov, "Report on an Artificial Satellite of the Earth" and also forwarded materials from the USA on work being carried out in this field. The current development of a new product (the R-7 ICBM) makes it possible for us to speak of the possibility of developing in the near future an artificial satellite...It seems to me that in the present time there is the opportunity...for carrying out the initial exploratory work on a satellite and more detailed work on complex problems involved with this goal. We await your decision.

Korolev convinced Minister Ustinov that if the projected R-7 could lift an estimated H-bomb warhead of 5 tons, it could easily launch a satellite of 1.5 tons, but the satellite project had to be presented as a stage in the development of the ICBM. The request went all the way to the Soviet leader Georgiy M. Malenkov who approved the idea. It is interesting to note that Korolev not only passed on Tikhonravov's report, but also "forwarded materials from the USA on the work being carried out in this field." Quite what these materials were, we can only speculate. The Soviet intelligence on the US missile and space program was on the whole achieved by diligent scouring of the open sources such as technical journals and the press. However, espionage reports cannot be ruled out, even though the state of the American space program was practically non-existent at that time. But that was all to change the very next month.

★ ★ ★ ★ ★

Back in the US, Project ORBITER began in June 1954, when Frederick C. Durant III, president of the International Astronautical Federation (IAF), requested that the Office of Naval Research (ONR) set up a meeting to investigate the launching of a satellite. It was to be a joint army-navy project and because of the importance of Wernher von Braun in the project, the secret meeting was timed to coincide with a visit of his to Washington on June 25th. Durant was not only the IAF president but also a past president of the American Rocket Society and currently working for the scientific intelligence unit of the CIA. Also present at the meeting were the upper-atmosphere physicist Fred L. Whipple, S. Fred Singer from the original Silver Spring group, and Lt Commander George Hoover of the ONR's Air Branch.

Von Braun turned up to the meeting ready armed with an army proposal for the launching of a satellite using the existing Redstone missile he had developed at Huntsville as the first stage, with upper stages of clustered anti-aircraft Loki rockets. He confidently announced that he could put a 5-lb satellite into space fairly soon.

On August 3rd, Lt Commander Hoover visited the Redstone Arsenal in Huntsville with a superior officer to further confer on the joint services project. Von Braun had prepared a paper for the group entitled "The Minimum Satellite Vehicle Based Upon Components Available From Missile Development of the Army Ordnance Corps." All seemed to be going well for the joint army-navy program. The US Air Force was also proceeding nicely with its own, independent satellite program. It obtained Pentagon authorization to begin work on building a reconnaissance satellite with the endorsement of the chief of staff General Nathan F. Twining, the Strategic Air Command chief General LeMay, and the World War II combat hero and chairman of the Air Force Scientific Advisory Board, James Doolittle.

After a further ORBITER meeting in Washington on September 6th, 1954, three members of the team attended a meeting of the Upper Atmosphere Rocket Research Panel, also in Washington, where it is assumed they briefed the space scientists during the classified sessions. One of those present at the Panel meeting was Milton Rosen from the Naval Research Laboratory, an organization not to be confused with the Office of Naval Research. Milton Rosen, the chief engineer of the Viking missile program, would soon become a powerful competitor to the von Braun-led team.

Von Braun produced a formal proposal dated September 15th, 1954, in which he stated, "The establishment of a manmade satellite, no matter how humble, would be a scientific achievement of tremendous impact...it would be a blow to US prestige if we did not do it first." The humbleness of the satellite was von Braun's admission that his 5lb inert object would have no real scientific use. Because of this, the project soon acquired the code-name Project SLUG. The reasons for the small size of the satellite were twofold. Firstly, it would help keep down the cost of the project and not annoy the budget-conscious Eisenhower administration, and secondly, it would speed up the development program in order to beat the Russians. With the USAF ICBM programs receiving maximum priority in 1954, von Braun kept the cost down to an estimated $100 thousand for the next fiscal year. This would be financed by the ONR whilst the cost of the Redstone booster would be absorbed by the US Army Ordnance's missile program.

The Redstone missile would act as the satellite's first stage, with a cluster of twenty-four or thirty-six Loki rockets for its second stage, a cluster of six for its third stage, and a single Loki rocket, attached to the satellite payload, as its fourth stage. The small size of the satellite, only weighing 5lb, seemed to rule out any installation of a radio beacon. This was a fundamental flaw in von Braun's proposal. All his effort was directed at getting the object into space, with very little concern about what the object was. Without a radio beacon, how was its position to be determined? How would proof be obtained of it even reaching space and establishing orbit? No experiments could be carried out so its scientific justification would not be believed. It was nothing more than an inert object launched into space to win a race.

Attempts were made to rectify this oversight by creating ludi-crous ways of signalling its position. Von Braun's team suggested that the satellite should be a mechanically unfolding sphere about twenty inches in diameter, which would reflect the sun's light and be visible through powerful telescopes. Von Braun also proposed that the rocket be given an equatorial launch in order to achieve the maximum velocity from the rotation qf the earth itself. This would require a launch from a remote Pacific Island chain or from the US Navy's experimental missile ship, the USS *Norton Sound*. This latter method of launch had been used earlier to fire a navy Viking missile. With this proposal, von Braun expected the Pentagon to agree, never suspecting that there would soon be competition from another branch of the US Navy.

On October 4th, 1954, exactly three years to the day before the launch of *Sputnik*, the Special Committee for the International Geophysical Year (*Comité speciale de géophysique internationale*) approved the US-sponsored plan to launch satellites during the IGY. This proposal clearly surprised the Soviet delegation that up until then had only attended the Rome conference as silent observers. The Soviet Union had ignored the May 1954 deadline for submis-sions for participation in the International Geophysical Year (IGY), but the satellite proposal sent shockwaves back to the USSR Academy of Sciences. The Soviet delegation suddenly announced that it was finally going to join the IGY. The Academy created the long-winded Interdepartmental Commission for the Coordination and Control of Work in the Field of Organization and Accomplishment of Interplanetary Communications to deal with the IGY, under the chairmanship of Academician Leonid I. Sedov.

In November 1954, with both American and Soviet scientists starting to talk about the prospect of satellites during the IGY, it is remarkable to note that US Secretary of Defense Charles E. Wilson denied knowledge of any American scientists working on Earth satellites and stated that he would not be alarmed if the Soviets were the first to build one. At the same time, the ORBITER group was working on a schedule of launching four satellites, commencing with the first in September 1956. Despite Wilson's comments, the CIA were showing an interest in von Braun's work, appreciating the psychological warfare value as well as the scientific prestige which would accrue to the United States if von Braun launched the first satellite. RAND, taking a rather partisan view, advised against a minimal satellite that could alert the Soviets to the possibility of a military surveillance satellite, such as that being developed by the US Air Force.

With the failure of the RAF Canberra spy mission over Kapustin Yar and the increased frequency of air attacks on the periphery of the Soviet Union, another source of intelligence on the Soviet missile program was needed. The National Security Agency (NSA) was running a radar listening post at Kiyarbakir, Turkey, which was about six hundred miles from Kapustin Yar. Using antennae half the length of a football field, the radar could monitor the test missiles as they rose above the horizon. But what was needed was a new type of reconnaissance plane that could fly far higher than the Soviet jet fighters sent to intercept it.

On Valentine's Day, 1955, a special panel of experts, known as the Technological Capabilities Panel (TCP) reported back to President Eisenhower on possible ways for science and technology to be used to defeat the enemy. One sure-fire method was to be left out of the official report because of its extreme sensitivity. During their investigations, the panel had discovered a specially designed photographic aircraft that would fly at seventy thousand feet, way above Soviet radar, fighters, and surface-to-air missiles. The air force had rejected this powered glider designated the CL-282, even though its designer had had an amazing track record. Clarence "Kelly" Johnson was head of the Lockheed "Skunk Works" and had designed the P-38 fighter-bomber and the F-104 Starfighter. The glider would have a special long focal-length camera to photograph objects as small as a man in strips two hundred miles wide. The panel suggested that the USAF-rejected project be handed over to the CIA. Eisenhower agreed and the U-2 was born.

The TCP chairman was James Killian and the panel's recommendations came to be known as the Killian Report. The TCP had three projects, focussing on offensive capabilities, defense technology, and intelligence. The panel on intelligence consisted of Edwin Land and Allan Latham Jr. of Polaroid, lens designer James G. Baker and physicist Edward Purcell of Harvard, chemist Joseph P. Kennedy of Washington University, and John Tukey of Princeton University and Bell Telephone.

These experts recognized that any overflight of the Soviet Union by the CL-282 would clearly constitute a violation of international law, but a satellite, flying much higher, would not necessarily do so. Edwin Land questioned where a nation's "airspace" ended and "space" began. International law needed to be established over the definition of national territory as it extended up into the atmosphere. Land's committee recommended to Eisenhower that the United States should develop a small artificial earth satellite to establish the right of "freedom of space" for larger, military intelligence satellites.

There needed to be a distinction between "national airspace" and "international space." Once legally established, the United States could then use the right to international space to its advantage by directing spy satellites over the Soviet Union land mass. Although the panel concluded that current 1955 technology was incapable of realizing the dream, and that photoreconnaissance through the CL-282 and the air force high altitude balloon program would provide better short-term results, research should be accelerated to produce a civilian satellite which would clear the path for the later military surveillance satellite. In part five of the classified Killian Report, the section dealing with intelligence matters, the panel assessed the potential uses of satellites:

We are convinced that small, inexpensive satellites can have several important uses related to intelligence. First, careful observations of the motions of the satellites would give, directly, much information about friction in the extreme upper air, and about fine details of the shape and gravitational field of the earth. This information is wanted for the design of large satellites for intelligence purposes. Second, even a lightweight satellite might be constructed to open into an extended structure (resembling a bedspring) which could be an effective reflector for very-high-frequency radio and radar signals. Conceivably, this could have applications in communication and intelligence. Third, the new prestige that the world will accord to the nation first to launch an artificial earth satellite would better go to the US than to the USSR. And although it is clear that a very small satellite cannot serve as a useful carrier for reconnaissance apparatus, cameras, etc., it can serve ideally to explore or establish the principle that space, outside our atmosphere, is open to all...Intelligence applications warrant an immediate program leading to very small artificial satellites in orbits around the earth. Construction of large surveillance satellites must be deferred until adequate solutions are found to some extraordinary technical problems in the information gathering and reporting system and its power supply; and should wait upon development of the ICBM rocket-propulsion system.

Exactly a month later, the US National Committee on the International Geophysical Year of the National Academy of Sciences presented a recommendation to Alan Waterman, the Director of the National Science Foundation, that a scientific satellite should be developed as part of the IGY. This was forwarded on to the Assistant Secretary of Defense for Research and Development, Donald Quarles, who having received the TCP's Killian Report, found the answer to the problem of finding a small satellite to pave the way for the larger intelligence ones.

It seems that the National Academy of Sciences' recommendation was totally independent of the classified TCP's findings. If this is the case, then it is a fine example of a fortuitous coincidence.

Quarles asked Waterman to formally suggest the idea of a small scientific satellite to the National Security Council, which Waterman
subsequently did in a letter to Robert Murphy, Deputy Under
Secretary of State. On March 22nd, Murphy met with Waterman,
Detlev Bronk, the President of the National Academy of Sciences,
and Lloyd Berkner, one of the original group who had created the
IGY back in Van Allen's house at Silver Spring five years before.

As the civilian satellite project started to materialize, the military surveillance one was well underway. The USAF had just for-
malized its requirement under the designation Weapons Systems
117L or WS-117L. As an offshoot of the RAND Project FEED BACK,
the WS-117L was formally authorized in a document entitled
"System Requirement (No.5) for an Advanced Reconnaissance
System." The USAF's objective was to provide continuous surveillance of "preselected areas of the earth" to determine the enemy
capability of launching air and missile strikes. The satellite would
be incorporated into the second stage of a rocket boosted by an Atlas
missile as its first stage. The Kodak film would be automatically
processed on board the satellite, then scanned electronically before
being transmitted down to earth like a television broadcast.

The technology required was too great for the time. The number
of video images that could be transmitted was restricted by too
many factors to be feasible in the near future. There was the short
time that the satellite would be within radio range of the ground-
tracking stations as it orbited the earth. In order to compensate for
this shortage of data, the satellite would have to stay up in space
for longer. To overcome the atmospheric drag on the satellite, which
would cause the object to lose altitude, it would have to be placed
in a higher orbit of at least three hundred miles, but this would
affect the detail quality of the pictures. The amount of film carried
on board would also have to be increased. The answer was to
develop the technology to photograph, scan, and transmit at far
faster rates.

Assistant Secretary Quarles attempted to get the air force to join
the army-navy joint project, but the USAF representative did his

best to scupper the whole plan. Major General Bernard A. Schriever wanted nothing to do with the proposed scientific plan either. He saw it as a mighty hindrance to the top-priority ICBM program and likely to slow it down at a time when the Soviet threat was at its greatest. Schriever was also unhappy about working on the concept with the "enemy," i.e., von Braun's team at Huntsville. Again perceiving a threat to the ICBM program, Schriever even prevented his own service's "World Series" proposal getting the support it needed from the Air Research and Development Command. This involved a modified Aerobee rocket on top of an Atlas ballistic missile and was suggested by Dr Ernst Steinhoff—one of von Braun's former colleagues at Peenemünde, now working at the USAF research center at Holloman AFB.

On April 15th, 1955, forty of the Peenemünde Germans and their families gathered in the auditorium of the Huntsville High School where two federal district judges administered the oath of citizenship. At last, von Braun and his team became US citizens, and as such were now fully cleared to work on top-secret material for the satellite program.

Quarles had received the von Braun army-navy proposal, Project ORBITER, now more realistically budgeted at $8.5 million, from the Secretary of the Navy in March, only to receive a second navy proposal in mid-April. This time it was a Viking-based system from Milton Rosen's Naval Research Laboratory. After having learned of Project ORBITER from the Upper Atmosphere Research Panel meeting the previous September, Rosen decided to provide a more efficient system with a definite scientific purpose, at a cheaper price of $7.5 million.

Since Rosen's NRL was responsible to the Office of Naval Research (ONR), von Braun's partner in the army-navy ORBITER proposal, the navy had to present the Viking system as a "backup for ORBITER and a possible second phase of a scientific satellite program." It was not presented as a competitor but in fact it was. The USAF "World Series" Atlas-Aerobee promised a larger payload but a projected cost of $50-100 million at some indeterminate time in the future, depending on the whim of Major General Schriever. Quarles dealt with the sudden appearance of two extra competitor bids for the possible IGY satellite contract by handing the problem over to his Committee on General Sciences for further study.

On May 4th, the committee reported back with its conclusions, and recommended that all projects should be pursued. ORBITER was selected as the official program, with a reduced budget of $6 million, and the Viking project was to be a similarly reduced $5.5 million backup, with the USAF project as a $1.25 million design study. The recommendations would next be put to the National Security Council scheduled for May 20th.

Joseph Kaplan of the National Academy of Sciences had a definite preference for the NRL (Viking) project, writing to his Director the next day that the US National Committee for the IGY should openly sponsor Rosen's NRL system. The combination of Viking and Aerobee would, he said, "clearly establish...the civilian character of the endeavor." The ORBITER, with its Redstone missile was, according to Kaplan, a "German V-2 development." The Nazi connection would be a liability, whereas the NRL system presented the best case for international science.

On May 16th, the *Washington Post* reported that Moscow Radio had announced the creation of a space flight commission that would place a space laboratory in orbit. Kaplan wrote to Alan Waterman of the National Science Foundation, stressing that time was obviously running out for the US scientific satellite project. Funding for the project would have to be agreed on by July 1955, or there would be no opportunity to arrange a satellite launch during the IGY period.

A few days before this, the man who acted as President Eisenhower's psychological warfare advisor, Nelson A. Rockefeller, urged the NSC to act immediately. Rockefeller was very concerned about the lack of perception among decision makers that there was a race on to be first into space. He wrote a lengthy memorandum cautioning of the "costly consequences of allowing the Russian initiative to outrun ours through an achievement that will symbolize scientific and technological advancement to peoples everywhere…the stake of prestige that is involved makes this a race that we cannot afford to lose."

Rockefeller was also aware of the dangers of an American success. The Soviets would employ vigorous propaganda against the achievement, so this must be countered by ensuring that the launch was as legitimate as possible. This meant having a satellite that was not merely an inert object, but a sophisticated piece of instrumentation superior to anything that the Soviets would subsequently put up. Rockefeller foresaw problems if America was first into space, but only with a satellite that did nothing: the achievement would be quickly discounted if the Soviets followed with a scientific satellite carrying a whole laboratory of experiments.

The launching of any US satellite would draw Soviet accusations of militarizing space, so Rockefeller recommended that the satellite should be put under the auspices of the IGY and its data shared with the international scientific community as a clear sign that the program was a peaceful venture and not a military one. Rockefeller's memorandum to the NSC stressed the psychological warfare dimension of the satellite project, following in the footsteps of the initial RAND report of May 1946, and the numerous warnings of Wernher von Braun.

The National Security Council met on May 20th, 1955, to approve a draft policy document known as NSC 5520 entitled "US Scientific Satellite Program." Rockefeller's memorandum was handed out to the members present, including President Eisenhower, Vice President Richard Nixon, Secretary of State John Foster Dulles, Secretary of Defense Charles Wilson, UN Ambassador Henry Cabot Lodge, and Allen Dulles from the CIA. The secret policy paper listed various reasons why the satellite project should be approved:

From a military standpoint, the Joint Chiefs of Staff have stated their belief that intelligence applications strongly warrant the construction of a large surveillance satellite. While a small scientific satellite cannot carry surveillance equipment and therefore will have no direct intelligence potential, it does represent a technological step toward the achievement of the large surveillance satellite, and will be helpful to this end so long as the small scientific satellite program does not impede development of the large surveillance satellite.

Furthermore, a small scientific satellite will provide a test of the principle of "Freedom of Space." The implications of this principle are being studied within the Executive Branch. However, preliminary studies indicate that there is no obstacle under international law to the launching of such a satellite. It should be emphasized that a satellite would constitute no active military offensive threat to any country over which it may pass. Although a large satellite might conceivably serve to launch a guided missile at a ground target, it will always be a poor choice for the purpose. A bomb could not be dropped from a satellite on a target below, because anything dropped from a satellite would simply continue alongside in the orbit.

Of the NSC members present, the UN Ambassador Henry Cabot Lodge was just as concerned as Rockefeller about world reaction to a successful American launch. It could easily be perceived as the United States militarizing space and world opinion would definitely be against that. Quarles reassured Lodge that the auspices of the IGY would shield the United States from such criticism. Lodge nevertheless cautioned that the public announcement of the plan would have to be handled very carefully.

Rockefeller proposed that the announcement should be made at the United Nations in accordance with the paper's recommendations concerning the peaceful nature of the program. CIA director Allen Dulles emphasized the importance of the project in the future gathering of intelligence. The document itself had recalled the Killian Report's recommendations as well as noting that top Soviet scientists were believed to be working on a satellite program. Eisenhower wound the meeting up by asking if anyone had any objections to the plan. There were none. The NSC endorsed the policy paper and the next day Eisenhower officially approved the satellite plan.

Soon after, Donald Quarles was put in charge of starting the scientific satellite program. In order to investigate the suspiciously low cost estimates he had previously been given by two of the three teams bidding for the contract, and to help select the satellite and its method of launch, Quarles created an ad hoc Advisory Group on Special Capabilities. As the chairman of this top-secret group, Quarles chose Homer Joe Stewart, an engineer-administrator from the Jet Propulsion Laboratory in Pasadena who also sat on the Air Force Scientific Advisory Board. The group became known as the Stewart Committee and consisted of eight members, with two being nominated by each service and two by Quarles' office at the Department of Defense.

The question of "Freedom of Space" was dealt with further in a progress report known as NSC 5522 dated June 8th, 1955. The Departments of State, Treasury, Defense and Justice commented:

> Any unilateral statement by the US concerning the freedom of outer space is unnecessary. It is clear that the jurisdiction of a state over the airspace above the territory is limited and that the operation of an artificial satellite in outer space would not be in violation of international law. State and Justice point out that by the convention of international civil aviation of 1944 (to which the US is a party, but the USSR is not) and by customary law every State has exclusive sovereignty "over the airspace above its territory." However, airspace ends with the atmosphere. There can be no recognition that sovereignty extends into airless space beyond the atmosphere.

How did anyone know the exact point where the atmosphere ended and space began? The edge of the atmosphere was not a clearly defined boundary until satellites actually managed to get up there and study it. The government lawyers advised that it was best not to publicly speculate on this question, as this might raise an awareness that did not exist at the time.

The CIA's opinion is worth repeating in full:

The psychological warfare value of launching the first earth satellite makes its prompt development of great interest to the intelligence community and may make it a crucial event in sustaining the international prestige of the United States. There is an increasing amount of evidence that the Soviet Union is placing more and more emphasis on the successful launching of the satellite. Press and radio statements since September 1954 have indicated a growing scientific effort directed toward the successful launching of the first satellite. Evidently the Soviet Union has concluded that their satellite program can contribute enough prestige of Cold War value or knowledge of military value to justify the diversion of the necessary skills, scarce material, and labor from immediate military production.

If the Soviet effort should prove successful before a similar United States effort, there is no doubt but that their propaganda would capitalize on the theme of the scientific and industrial superiority of the communist system. The successful launching of the first satellite will undoubtedly be an event comparable to the first release of nuclear energy in the world's scientific community, and will undoubtedly receive comparable publicity throughout the world. Public opinion in both neutral and allied states will be centered on the satellite's development.

For centuries scientists and laymen have dreamed of exploring outer space. The first successful penetration of space will probably be the small satellite vehicle recommended by the TCP. The nation that first accomplishes this feat will gain incalculable prestige and recognition throughout the world. The United States' reputation as the scientific and industrial leader of the world has been sharply challenged by Soviet progress and claims. There is little doubt but that the Soviet Union would like to surpass our scientific and industrial reputation in order to further her influence over neutral states and to shake the confidence of states allied with the United States. If the Soviet Union's scientists, technicians, and industrialists were apparently to surpass the United States and first explore outer space, her propaganda machine would have sensational and convincing evidence of Soviet superiority.

If the United States successfully launches the first satellite, it is most important that this be done with unquestionable peaceful intent. The Soviet Union will undoubtedly attempt to attach hostile motivation to this development in order to cover her own inability to win this race. To minimize the effectiveness of Soviet accusations, the satellite should be launched in an atmosphere of international goodwill and common scientific interest. For this reason, the CIA strongly concurs in the Department of Defense's suggestion that a civilian agency such as the US National Committee of the IGY supervise its development and that an effort be made to release some of the knowledge to the international scientific community.

This small scientific vehicle is also a necessary step in the development of a larger satellite that could possibly provide early warning information through continuous surveillance of the USSR. A future satellite that could directly collect intelligence data would be of great interest to the intelligence community. The Department of Defense has consulted with the Agency and we are aware of their recommendations, which have our full concurrence and strong support.

The reality of the Soviet space program at the time was slightly different. On July 16th, Tikhonravov published his latest study on artificial satellites, developing some of the ideas in his original May 1954 document. He recommended a workforce of between seventy and eighty people to carry out the designing and building of the satellite. The Chief Designer, Korolev, added his comments that the "creation of a satellite would have enormous political significance as evidence of the high development level of our country's technology" but that didn't stop him reducing the proposed workforce level by half. The Soviet space program was still in the stalls.

The CIA had a very great interest in the question of "Freedom of Space." It would make their job a lot easier. The man responsible for liaising with Alan Waterman of the National Science Foundation was Richard Bissell, then in charge of the high altitude reconnaissance glider designed by Kelly Johnson's "Skunk Works," code-named AQUATONE. Its planned overflights of the Soviet Union would violate international law, whereas a surveillance satellite in international space would not.

President Eisenhower was also concerned about the development of the AQUATONE aircraft and its illegal operations. In an attempt to solve the problem of overflight, Ike presented an "Open Skies" proposal at the Geneva Summit of the Big Four on July 21st, 1955. The surprise announcement was apparently aimed at removing the threat of surprise attack. The president reasoned that if aerial reconnaissance were allowed for the purposes of monitoring military sites, both sides would have enough information to eliminate any unnecessary fears of nuclear mobilization. "Open Skies" would be the beginning of "effective inspection and disarmament." If the Soviets agreed to "Open Skies," Eisenhower would have free reign to use the AQUATONE glider with its advanced photoreconnaissance capabilities.

However, the Soviets did not like the plan. Prime Minister Nicolai Bulganin argued that "if the military in both the United States and the Soviet Union were to learn all about each other's armed forces, they would automatically begin agitation for increases in their own; therefore the Open Skies would increase rather than decrease armaments."

Khrushchev put it rather more bluntly: "This is a very bad idea. It is nothing more than a spy system."

A few days later, Kelly Johnson delivered the first AQUATONE glider to the secret Nevada test site that would become well known to conspiracy theorists as Area 51. To conceal the aircraft's mission, the former CL-282 was put on the US Air Force's books as a utility aircraft. Since the designations U-1 and U-3 were already allocated, the plane became the U-2.

President Eisenhower returned from Geneva with his "Open Skies" plan rejected by the Soviets, and this may have prompted him to authorize the press announcement of the US satellite program. Intelligence reports also suggested that the Soviet Union was about to make a similar announcement about a Soviet program. The US announcement was delivered not at the United Nations, as UN Ambassador Henry Cabot Lodge had recommended at the June meeting of the National Security Council, but at the White House. Eisenhower's Press Secretary James Hagerty arranged a secret preliminary press briefing on July 28th.

Locking them into the White House press room, and forbidding them to report their stories until the next day, Hagerty told correspondents: "The President has approved plans by this country for going ahead with the launching of small unmanned earth-circling satellites as part of the United States participation in the International Geophysical Year...This program will, for the first time in history, enable scientists throughout the world to make sustained observations in the regions beyond the earth's atmosphere. The President expressed personal gratification that the American program will provide scientists of all nations this important and unique opportunity for the advancement of science."

When Hagerty was asked by one reporter if the satellite project was linked to military satellites planned by the Pentagon, he replied: "The only connection the Department of Defense will have with this project is actually getting these satellites up into space."

The sudden announcement of the plan at the White House was given in advance of an important scientific conference. The reporters had been sworn to secrecy until the next day so that Marcel Nicolet, the Secretary of the special committee for IGY, or CSAGI (Comité Spécial de l'Année Géophysique Internationale), could be notified of the decision first. The international scientific community seemed to use very old-fashioned methods of communication, since a hand-written letter was carried from Washington to New York that night by an assistant to Waterman, who then handed it over to a friend bound for London, where it was passed on to a Dr Brook who would deliver it to Nicolet on the 29th.

At the opening of the Sixth International Astronautical Federation Congress in Copenhagen on August 2nd, 1955, when IAF President Fred C. Durrant III announced Eisenhower's plans to the assembled scientists, it was the first time that the Soviet delegates, Academician Leonid Sedov and astronomy journal editor Kirill F. Ogorodnikov, had attended. Not to be outdone, Sedov immediately arranged for a press conference the same day at the Soviet Embassy in Copenhagen, where he announced that the Soviet program to launch a satellite would be within eighteen months.

The figure rang immediate alarm bells back in Washington as well as Moscow, since the Council of Ministers had not even issued a decree authorizing the satellite's development yet. A Soviet launch in eighteen months' time would be six months before the official start of the IGY—on July 1st, 1957. There was a possibility that Sedov had meant the eighteen months that the IGY actually spanned, i.e. from July 1st, 1957, to December 31st, 1958, but the confusion had done its damage despite his withdrawal of the comment the next day: "From a technical point of view, it is possible to create a satellite of larger dimensions than that reported in the newspapers which we had the opportunity of scanning today. The realization of the Soviet project can be expected in the comparatively near future. I won't take it upon myself to name the date more precisely."

With Serov's comments, the race was on.

Von Braun and his Rocket Team at Huntsville were confident that their Redstone rocket would be chosen as the booster for the US scientific satellite. They had developed the Redstone in the early 1950s when the Army Ordnance ran out of the V-2s it had acquired from the Nordhausen Mittelwerk. It was a liquid fuel rocket about thirty percent more powerful than the V-2 and it weighed considerably less. The ORBITER proposal used the Redstone as the first stage with a clustering of Loki rockets for the upper stages. Quarles' Committee on General Sciences had already recommended the ORBITER as the main satellite launch system with the NRL Viking Aerobee as backup, but the decision was now to be made by the Stewart Committee. Its chairman, Homer J. Stewart, had already been involved in the early ORBITER planning when he had completed a study of von Braun's September 1954 proposal for W.H. Pickering, Director of the Jet Propulsion Laboratory, who wanted to get in on the Huntsville-ONR project. With the Committee chairman in his camp, von Braun felt confident that his proposal would succeed.

Then came an almighty shock. The committee ignored its chairman's advice and, on August 4th, 1955, it voted for the Naval Research Laboratory's proposal, presented by Milton Rosen.

The Stewart Committee

Werner von Braun was understandably irate at the Stewart Committee's decision. The Redstone Arsenal team were by far the most experienced rocket scientists in the United States. They had learned by making mistakes, from Peenemünde to Blizna, from White Sands to Huntsville and Cape Canaveral. The Naval Research Laboratory had none of that experience; they would have to go through the same evolutionary process and that would take time—valuable time that the United States did not have in the race against the Soviets. Von Braun even had the legendary Hermann Oberth on his team, having rescued him from obscurity. The Transylvanian theorist worked on highly classified projects to which, for national security reasons, he was denied access due to his foreign status. This strange situation, where he could not have access to his own work, did not prevent him developing further some of his earlier Nazi weapons such as the orbiting mirror than could either apply light or intense burning heat to any spot on the earth's surface. Von Braun would have welcomed the mirror weapon to provide illumination to the Stewart Committee.

The excuses given for not selecting the obvious favorite were numerous. The National Security Council document NSC 5520 explicitly stated that the scientific satellite was not to interfere with the intercontinental and intermediate range ballistic missile programs. The Redstone booster was to be used as the basis of an intermediate range ballistic missile (IRBM) designated the Jupiter, thus effectively ruling it out. The Atlas component of the USAF proposal mean that would have been denied. That left just the Viking Aerobee of the NRL. There was also a problem with von Braun's proposal being dependent on the Loki rockets for the upper stages. There was nothing wrong with his Redstone first stage, but plenty wrong with the Lokis. The Loki solid rockets did not have a great reliability record. The more Lokis used in the clustering, the greater the chance of some of them failing. Von Braun and his team had arranged the Lokis in three stages of thirty, six, and one. All it needed was a single failure and the satellite would not be able to reach orbit.

Despite reassurances from the Loki contractor, Aerophysics, the ORBITER proposal was definitely hampered by its choice of second, third, and fourth stage rockets. Back in April, before the establishment of his committee, when Homer Stewart had reported to JPL's director Pickering on the ORBITER, he suggested that the Loki rockets be replaced with the six-inch diameter Sergeants. Pickering passed on the suggestion to von Braun at a meeting in Pasadena shortly afterwards. Eleven Sergeants could be clustered round the missile's second stage, followed by three at the third stage, and one at the fourth. With less to go wrong, the arrangement would have been preferable. Von Braun did mention it to the Stewart Committee when the ORBITER pitch was made, but proposed the as-yet-untested seven-inch diameter Sergeants arranged in a different configuration. The alteration of Stewart's recommendation may have been an unwise move, signalling unwillingness on the part of von Braun and his team to take outside advice.

But by far the greatest criticism of von Braun's ORBITER was the pathetic size and content of the satellite. At a mere 5lb and with no real capacity for IGY experiments, the satellite was still basically the inert sphere it had been when nicknamed Project SLUG. Von Braun admitted in his proposal that the ORBITER satellite could also carry the NRL competitor's radio system, or an army-designed radar reflector that would modulate the reflected signal using a small power source on board. The most that the ORBITER proposal could do was allow itself to be tracked in orbit as a method of studying the Earth's gravitational field and the extreme outer atmosphere.

This had to be compared with the 34lb heavily instrumented satellite offered by the NRL.

There was a miniature electron-beam vacuum tube and a crys-
tal-controlled radio receiver with a hearing-aid-type amplifier to
respond to instructions sent from earth and telemeter equipment
to transmit information back to the ground stations. There were two
Lyman-alpha detectors on board to measure the intensity of the
ultraviolet radiation emitted by atomic hydrogen from interstellar
space and radiation coming from closer to home, the sun. Data
would be supplied on the particle streams ejected from the sun,
which scientists believed affected the intensity of cosmic rays, the
aurora, and magnetic storms. Another experiment would be served
by a magnetometer that would detect the presence of a hypothet-
ical ring current encircling the earth and gauge the intensity of the
planet's magnetic field. The altitude of the satellite would be determined by the miniature electron beam vacuum tube that would
measure the angle between an axis in the satellite and the earth's
magnetic field. As far as the satellite was concerned, there was no
comparison between the NRL's scientific attractiveness and the
feeble 5lb sphere of the ORBITER.

The competition between the rocket boosters was another matter.
The NRL had actually offered two possible launchers in its proposal.
Both were based on the M-10 Viking, a smaller missile than the
Redstone with a diameter of four feet, a length of forty feet and a dry
weight of 2,250lb. Designed by the Glenn L. Martin Company, the M-
10 Viking was a modification of the US Navy's sounding rocket which
had reached altitudes of over a hundred miles in recent tests. One
configuration proposed was an M-10 Viking with two solid fuel upper
stages designed by the Atlantic Research Corporation. It would place
the satellite into an orbit at a perigee of 216 miles. The second configuration was the M-15, which would reach a perigee of 303 miles.
It consisted of an M-10 Viking with an Aerobee-Hi liquid propellant
second stage and a solid third stage. The Aerobee-Hi was untested;
its first flight was scheduled for the next month. Its manufacturer,
the Aerojet General Corporation, was ready to supply the rockets.

The Stewart Committee chose the M-15 configuration, even though the NRL had warned that it would take six months longer than the M-10 to develop. The three-stage design was viewed as superior to the four-stage clustered arrangement, as were the two liquid fuel stages. It is ironic that the German originators of liquid fuel propellants in rockets should have become so dependent on solid fuel upper stages. Von Braun criticized the liquid fuel to be used in the Aerobee-Hi stage, unsymmetrical dimethylhydrazine (UDMH), which had not yet been tested. Even the solid propellant in the third stage, ammonium perchlorate in a polyvinyl chloride matrix, had not yet been tested in large motors.

Quarles, faced with the Stewart Committee recommendation, deliberated and passed the buck on to his Policy Council. After listening to the arguments of its army and navy members, the Council voted to postpone the final decision for two weeks. Quarles faced an angry protest from von Braun's superior, Major General Leslie Simon of the US Army Ordnance Corps. Bringing up the Soviet threat, Simon attempted to sway the decision. He offered two new proposals.

The first suggestion was an upgraded Redstone first stage using North American Aviation's 135,000lb thrust engine in place of its 75,000lb one (compared with the Viking's 30,000lb). This would allow a satellite weighing 162lbs to be put into orbit with a perigee of 218 miles—more than enough to carry a whole range of instrumentation and experiments. With this scheme Simon promised a launch scheduled for December 1956, seven months before the IGY started and in direct competition with the Soviet date possibly suggested by Sedov.

However, Simon proposed that the best solution would be to allow the improved Redstone to place the satellite into orbit and have it carry instrumentation provided by the NRL. He was critical of the unrealistic development times of the NRL, but so confident of this solution that he claimed the improved Redstone would be able to put a 100lb payload on the moon. Quarles gave the ORBITER team a second chance and ordered the Stewart Committee to review von Braun's proposal. On hearing this, Rosen's team at the NRL were incensed that their fairly won victory was being snatched away from them and petitioned to get their case heard a second time too.

Quarles more or less accepted Simon's offer and overturned the Stewart Committee's recommendation. He would give the US Army $20 million to run the program and provide the booster. The NRL would provide the satellite and the tracking system with the navy funding the development of the Aerobee-Hi as an alternative second stage. The US Air Force would work on a backup using elements of its military satellite program, and provide launch support. Quarles managed to get the Research and Development Policy Council to agree to his tri-service compromise, conditional upon sanction by the Stewart Committee. They were to report in a week on the various improvements that both sides had made to their original presentations. Quarles' surprise compromise can be explained by his desire to sort out the satellite problem on his last day in the job. The Assistant Secretary of Defense for Research and Development was about to become the new Secretary of the Air Force.

The main improvements to the NRL's first presentation to the Stewart Committee were the availability of cost estimates at $12 million and an adjustment in the third stage, with a Sergeant-type rocket as suggested by the committee. This necessitated a reduction in the size of the satellite to 21.5lbs unless a new T-65 rocket from the Thiokol Chemical Corporation was used instead. Rosen was confident that the satellite could be launched within the 18-month period of the IGY. Rosen backed this up with letters of guaranteed delivery dates from his subcontractors.

Von Braun's second presentation failed to persuade the Stewart Committee members that the ORBITER with an improved Redstone was a better proposition than had already been offered. The air force had agreed to supply the massive 135,000lb engines in June 1957, and confirmed that they would not interfere with priority military programs. The development time added for the improved engines would rule out any advantage gain in increased payload if the Soviets were already in space. Also, the June 1957 delivery date made a mockery of Major General Simon's December 1956 early launch date. As far as this was concerned, the Stewart Committee believed the date was unrealistic anyway.

The Committee was also very sceptical about the ability of three services to work together in Quarles' compromise. It was almost too fairytale in its expectation. The realities of inter-service rivalries were active within the Stewart Committee itself, with members selected for their particular service loyalties or relationships. There was another reason why the Viking system was preferable to the Redstone. The NSC 5520 wanted scientific data to be made freely available to international scientists, as part of its cover up for the secret development of a surveillance satellite. This effectively ruled out the Redstone rocket, since it was part of the United States' missile force and therefore classified. There was no way that foreign scientists would be allowed access to Huntsville as part of the IGY spirit of scientific brotherhood.

When the Stewart Committee ignored Quarles' compromise and confirmed their selection of the NRL's Viking project, von Braun and the Huntsville team went truly ballistic. The day after the Stewart Committee's controversial decision, Homer Stewart travelled down to Huntsville and confided to von Braun, "We have pulled a boner." Although he had personally voted for the ORBITER, Stewart was to be blamed for the Committee's majority decision. The reporter Jack Anderson, who was preparing to write a column for Drew Pearson in his absence, phoned Dr Stewart at his home and asked for his explanation for the decision. Previously, Stewart had denied to the reporters that he was even the chairman of the secret committee that bore his name, so he was unlikely to reveal anything further. Anderson duly blamed him for selecting VANGUARD and someone placed a full-page ad in the *New York Times* recommending that he should be hanged. Feelings were running high in Huntsville.

Von Braun and the Army Ordnance officers believed that the rejection of ORBITER had nothing to do with the Redstone, Lokis, or the 5lb satellite. They believed that the truth behind the controversial decision was plain old racism. Von Braun's rocket team at

Huntsville was still perceived as the Nazi team from Peenemünde. The IGY wanted a scientific satellite that was not identified with war. The Redstone booster was a ballistic missile with an intermediate range capability within the United States' arsenal. It was clearly identified with war, having been developed from the V-2 missile. But this was nothing compared with the war-like association of the design team. Although now registered as US citizens, the Huntsville team was still looked upon as Germans with very undesirable pasts. There were many Americans who still did not trust their motives, and even more who had a never-ending hatred of them born out of personal wartime memories.

Was the navy project picked because it had no war-like appearance to the IGY Committee? No, it was after all a military rocket developed for the navy, so that argument has to be dismissed. Was the Viking picked because it was less military than the Redstone? Perhaps, but the Germans believed that their project had been rejected because the US government wanted an American satellite launched on an American rocket by an American team. Their American citizenship was meaningless. In fact, their ceremony in Huntsville a few months before now looked like a desperate attempt to legally qualify them in time for the IGY selection. The final snub to von Braun and his team came with the decision that their ORBITER project was not even commissioned as the backup. Von Braun was told to abandon all plans for a satellite launch vehicle.

On August 30th, 1955, Tikhonravov and Korolev attended a secret meeting at the offices of the Academy of Sciences. They were there to report on the latest developments in the Soviet space program. Korolev informed the Academicians that the R-7 intercontinental ballistic missile could easily be converted into a satellite vehicle. He proposed that the Soviet Union should embark on a series of launches, including one with animals. As for the timetable, Korolev stated, "We hope to begin the first launches in April-July 1957...before the start of the International Geophysical Year." This was in clear violation of the international agreement to launch during the IGY. If the Americans were honorable enough to stick to the rules and launch between July 1st, 1957 and December 31st, 1958, then the Soviets would launch earlier in order to secure first place in the space race. This appealed to the Academy of Sciences and they elected Mstislav V. Keldysh to be chairman of the new commission to be set up to achieve this Soviet victory, with Korolev and Tikhonravov serving as his deputies.

Because of its secret status in the security-conscious Soviet Union, the commission did not attend the annual CSAGI assembly in Brussels in September. The US IGY Committee prepared a major presentation of its own satellite program in the hope that some of the leading Soviet rocket experts would attend, but the solitary Soviet delegate kept very quiet indeed. This may have been because the Soviet satellite program was still only a proposal. It was not until January 30th, 1956, that the USSR Council of Ministers finally issued decree number 149-88ss, a document approving the launch in 1957 of a satellite designated OBJECT D. This satellite was based on Tikhonravov's design and was to be manufactured by the OKB-1, the same design bureau that produced Korolev's R-7 ICBM. It would weigh up to 3000lbs with over 600lbs being made up of scientific experiments. It would dwarf the NRL's satellite.

In February, Krushchev made his first visit to the OKB-1 with Bulganin and Molotov, accompanied by the Minister of Defense Industries, Dimitriy Ustinov. Up until then, the Soviet missile program had been run by people like Ustinov, with very little Party control. Now, with this high level state visit by Party members, Korolev and Tikhonravov saw the opportunity to promote the Soviet space program directly to the leadership. Taking them on a guided tour of the institute, Korolev showed them a full-scale model of the R-7 ICBM, which stunned them into silence with its sheer size. After a short presentation on its capabilities, Korolev chose his moment very carefully.

"Nikita Sergeyevich, we want to introduce you to an application of our rockets for research into the higher layers of the atmosphere and for experiments outside the atmosphere."

Khrushchev responded with polite interest but the rest of the Presidium members were showing signs of tiredness and, worst of all, boredom with the proceedings. Sensing this, Korolev acted fast, showing them a model of the OBJECT D satellite, and explaining that finally the dream of the great Tsiolkovsky would be realized with the help of the R-7. Korolev compared the mighty Soviet missile with that of the Americans' "skinny" Viking missile, and warned the visiting dignitaries about the advanced stage of the US space program. The far more powerful R-7 would be able to launch a satellite much larger than anything the Americans could put up into space.

Korolev played his final card by declaring that the costs for such a satellite project would be meager since the main expenditure would be the rocket booster and that was already absorbed into the R-7 program. OBJECT D was so named because it was the fifth type of payload to be carried on an R-7. OBJECTS A, B, V, and G (in the Russian Cyrillic alphabet) were nuclear warheads, so all that was required, Korolev explained, was the simple replacement of a warhead with a satellite. Khrushchev hesitated, then when he was assured that the satellite program would not interfere with the R-7 ICBM program, he gave his assent.

"If the main task doesn't suffer, do it."

The *National Geographic* magazine ran an article on the newly named VANGUARD project from the Naval Research Laboratory and the US IGY Committee. It referred to VANGUARD as "history's first artificial earth-circling satellite." Down in Huntsville, von Braun and his team doubted whether the Soviets could now be beaten. In a major restructuring, the Army Ballistic Missile Agency (ABMA) was established at Redstone under the command of Major General John B. Medaris. His unit absorbed von Braun's team and the Redstone missile, but despite being at the receiving end of a V-2 attack in France during World War II, the new commander held the Germans in highest regard. He was not about to allow Huntsville to miss out on the space race. On April 23rd, 1956, the ABMA informed the Office of the Secretary of Defense that a Jupiter-C missile could be fired in an attempt to launch a satellite into orbit as early as January 1957. ABMA proposed that the Department of Defense should consider this as an alternative to the navy VANGUARD program.

The Jupiter-C missile was in effect a Redstone with an experimental re-entry nose cone designed for use in the Jupiter missile. There was an urgent need for a nose cone which could survive the intense heat of re-entering the earth's atmosphere, without melting away and destroying the nuclear warhead prematurely. Von Braun's rocket team had been given the nose cone testing task, and this had kept alive their hopes of getting into space.

These hopes were rapidly quashed when the Assistant Secretary of Defense informed ABMA that it was not to initiate any program to launch a satellite using Redstone or the new Jupiter missiles. General Medaris took his case to the Deputy Secretary of State Herbert Hoover who was in charge of the Operations Coordinating Board. This body was responsible for overseeing the implementation of the NSC 5520 decision to launch a small scientific satellite. On May 28th, Assistant Secretary Hoover contacted General Andy Goodpaster, Eisenhower's staff secretary, about the problem. The latter recorded:

The best estimate is that the present project would not be ready until the end of '57 at the earliest and probably well into '58. Redstone had a project well advanced when the new one was set up. At minimal expense ($2-$5 million) they could have a satellite ready for firing by the end of 1956 or January 1957. The Redstone project is one essentially of German scientists and it is American envy of them that has led to a duplicative project. I spoke to the President about this to see what would be the best way to act on the matter. He asked me to talk to Secretary Wilson. In the latter's absence, I talked to Secretary Robertson today and he said he would go into the matter fully and carefully to try to ascertain the facts.

Reuben B. Robertson Jr., the Deputy Secretary of Defense, referred the problem to E.V. Murphree, the DoD's Special Assistant for Guided Missiles. Also within the DoD, the replacement for Donald Quarles, Charles C. Furnas, was finding his job as Assistant Secretary of Defense for Research and Development a frustrating one. With Quarles' last-minute compromise rejected by the Stewart Committee and the VANGUARD project adopted as the official entry in the IGY, Furnas saw the estimated budget for the VANGUARD mushroom. From an unrealistic $12 million at the time of the second Stewart Committee presentations in August 1955, the figure had almost immediately doubled in September to $28.8 million. After a further massive hike to $63 million in April 1956, it was now, in June, estimated at $99.2 million. What made Furnas' job even more frustrating was the fact that up until his appointment as Assistant Secretary of Defense for Research and Development, he had been the Chancellor of the University of Buffalo and one of those members of the Stewart Committee who had voted for the ORBITER and von Braun.

On June 22nd, Furnas met with his old TCP committee chairman, Homer Stewart, at the Pentagon to receive an update on the VANGUARD/ORBITER controversy. Stewart reported that although Project VANGUARD was horrendously over budget and there were some minor setbacks, it was on a satisfactory schedule that should place a satellite into orbit within the IGY timeframe. He mentioned that there was a shortage of qualified technical experts at the Naval Research Laboratory who could coordinate the work of the subcontractors. As far as the abundance of qualified technical experts in Huntsville was concerned, Stewart made the following comments in his secret written report:

Redstone re-entry vehicle No.29, now scheduled for flying in January 1957, apparently will be technically capable of placing a 17lb payload consisting principally of radio beacons and Doppler-equipment in a two-hundred-mile orbit, even with a degradation in performance below the present design figures which might reasonably be expected, but without further margin. This capability will depend upon successful accomplishment of several developments, such as the new fuel in the Redstone booster, and the spinning cluster of fifteen solid propellant motors. The probability of success of this single flight cannot be reliably predicted now, but it would doubtless be less than fifty percent. In any case, such a single flight would not fulfill the nation's commitment for the International Geophysical Year because it would have to be made before the beginning of that period. Adequate tracking and observation for the scientific utilization of results would not be available at this time. Moreover, any announcement of such a flight or, worse, any leakage of information if no prior announcement were made, would seriously compromise the strong moral position internationally which the United States presently holds in the IGY due to its past frank and open acts and announcements as respect VANGUARD.

Stewart suggested that von Braun's Redstone could be used as a backup if the VANGUARD started to fall behind schedule. No matter what, the United States had to launch some kind of satellite before the end of the IGY. Therefore, although the Stewart Committee did not recommend the Redstone option be activated, it did not rule out the possibility of reconsidering it later. Clifford Furnas, although he supported the Huntsville team, was aware that the sudden priority switch from Viking to Redstone would adversely affect the morale of the NRL team and sour relations with both the national and international scientific communities. Furnas discussed the Stewart Committee's latest report with the Special Assistant for Guided Missiles, E.V. Murphree, and he reported back to the Deputy Secretary of Defense, Reuben Robertson, who replied to General Goodpaster's request for an investigation. The White House Staff Secretary then forwarded the conclusion of "no change" on to Eisenhower.

The President had already authorized the first overflight of the Soviet Union by the U-2 but bad weather had delayed the date until

July 4th. The U-2 was given the supposed role of civilian weather reconnaissance at the fictional "1st Weather Reconnaissance Squadron Provisional" at Wiesbaden in the Federal Republic of Germany. Its pilot, Hervey Stockman, took off from Wiesbaden and flew over Poznan, Poland, then headed for Leningrad, his main target. It was at that city's shipyards that the Soviet submarines were built. Along the way, major Soviet military airfields were photographed to count the number of Bison heavy bombers. These were the targets of the next day's overflight that took in Moscow. The U-2 pilot on this occasion was Carmen Vito who brought photographs of the Bison airframe plant at Fili and the Bison test facility at Ramenskoye airfield outside Moscow back to Wiesbaden.

As far as sites significant to the rocket program went, Vito photographed Korolev's missile plant at Kaliningrad and Glushko's engine plant at Khimki. The pictures were of stunning quality. The use of haze filters meant that even the pictures of Moscow were good, despite the heavy cloud that had been covering the city. The U-2s had gone undetected by the Soviet Union since there had been no public outcry. Or so Eisenhower and his CIA advisors believed until a Soviet letter of protest arrived on July 10th, specifying the route flown, the depth of the penetration into Soviet territory, and the actual length of time spent in Soviet airspace. The U-2s had been tracked but the Soviets had been unable to do anything about them, much to their embarrassment. General Goodpaster immediately grounded all U-2 flights until further notice. Although the U-2's photo-reconnaissance missions had been successful, its inability to avoid Soviet radar meant operations had to be curtailed. Although the enemy knew of its presence, they could not shoot it down because of its extreme altitude. This reinforced Eisenhower's desire to switch from overhead aerial photoreconnaissance to a satellite-based system.

The first Jupiter-C rocket, numbered RS-27, was launched from Cape Canaveral on September 20th, 1956. It was essentially a Redstone first stage with two extra solid-propellant rockets developed by the Jet Propulsion Lab at Cal Tech. The Jupiter-C flew 3,355 miles, reaching an altitude of 682 miles and a velocity of Mach 18. Because of the Department of Defense's directive that von Braun was not to attempt a space launch, the Jupiter-C carried a dummy payload as its fourth stage. The engine was filled with sand ballast instead of solid fuel in order to prevent an "accidental" boost of the fourth stage into orbit. Both Medaris and von Braun later claimed that Washington had been so fearful that the ABMA would go ahead with an unauthorized satellite launch that two observers were sent from the White House Bureau of the Budget to ensure this did not happen.

Science fiction writer Arthur C. Clarke believed that von Braun had been planning to launch a clandestine satellite on a Jupiter-C but his secret plan had been frustrated in time by the Department of Defense. Whether this was a different version of the same event or a different event all together, it certainly shows that there was a very strong distrust of von Braun. The White House did not want the Huntsville team beating the VANGUARD team into orbit. It was almost as if the race was between two "American" projects, ignoring the very powerful third competitor. The following month, *Time* magazine ran an article on the progress of the VANGUARD project with the following information on its rival:

> Secrecy always breeds rumors and a widespread rumor in the missile business was that the army hoped to toss a satellite into the sky ahead of Project VANGUARD which was administered by the navy. Leader of this dark plot, according to rumor, was famed Wernher von Braun, chief creator of the V-2...Unless stopped by higher authority, army missile men may try to beat the navy into space.

The *Time* article was not denied by ABMA and it was said that Medaris treasured the story, keeping clippings in his scrapbook. Another person to keep the clippings was Sergei Korolev. Von Braun's September 1956 launch would send alarm bells ringing at

NII-88. Korolev believed that it really was a failed attempt to launch a satellite. The KGB tried to convince him that this was not the case, but he refused to accept their intelligence assessment. Korolev firmly believed that he was in a race with the Americans, especially the Germans at Huntsville, and the Soviet satellite program was hitting problems. Static test firings of the R-7 engines were producing impulse times too low for the heavy OBJECT D satellite. Korolev would have to reduce its size in order to get back on schedule. The design of Tikhonravov's was too complicated for a first satellite launch. What was needed was a far simpler version of the one and a half ton OBJECT D.

Korolev's misconception of the threat from von Braun would have delighted the Huntsville team if they had known; at least someone, somewhere, was appreciating their talent. On November 26th, the Germans' morale hit an all-time low as the Secretary of Defense Charles Wilson created specific missile roles for each of the three services and the army came out the worst. The navy was given control over ship-based missiles, the air force had control over long-range missiles, and the army was limited to those missiles with a range of less than two hundred miles. The Jupiter missile deployment was assigned to the United States Air Force with ABMA now reduced to the role of a contract supplier. Medaris later recalled: "The psychological impact at Huntsville was devastating, not only at the Arsenal but in the town."

Exactly how Korolev reacted to this news is not known, but the effective neutering of his prime competitor von Braun would not have aided his own ambitions. Korolev needed to request permission to change the size and scientific capabilities of the satellite, so the threat from von Braun needed to be maintained in the eyes of the USSR Council of Ministers. Therefore, Korolev ignored the America restrictions placed on von Braun and emphasized the need for haste:

> The United States is conducting very intensive plans for launching an artificial Earth satellite. The most well known project

under the name "VANGUARD" uses a three-stage missile...the satellite proposed is a spherical container of fifty centimeters diameter and a mass of approximately ten kilograms. In September 1956 the USA attempted to launch a three-stage missile with a satellite from Patrick Base in the state of Florida which was kept secret. The Americans failed to launch the satellite...and the payload flew about 3,000 miles or approximately 4,800 kilometers. This flight was then publicized in the press as a national record. They emphasized that US rockets can fly higher and farther than all the rockets in the world, including Soviet rockets. From separate printed reports, it is known that the USA is preparing in the nearest months a new attempt to launch an artificial Earth satellite and is willing to pay any price to achieve this priority.

In typical Soviet fashion, Korolev chose to ignore the fact that von Braun's 3,355 miles was not just a national record, but an international one too. The Soviet Union was still yet to launch its first ICBM. Korolev's green light was received on February 15th, 1957, when the USSR Council of Ministers formally issued a new decree, No. 171-83ss, entitled "On Measures to Carry Out in the International Geophysical Year." The new satellite was called *Prostreishiy Sputnik* or "PS", the "simplest satellite." Two of them would be launched in April and May 1957, after one or two successful launches of the R-7 rocket. The reduced size of the *Sputnik* was announced on Radio Moscow in March, giving its weight at fifty kilograms. Korolev had not completely abandoned the OBJECT D—it would be rescheduled as *Sputnik 3*.

Not only was there a new satellite for his R-7 but also a new missile test site to launch it from. This was at Tyuratam, a remote region of southern Russia near the Aral Sea. It was designed to conduct ICBM tests over a three thousand-mile range across Siberia to the Kamchatka peninsula. It was classified top secret and the Soviets had managed to build it without Western knowledge until Eisenhower authorized the U-2 overflights to recommence. Flying over Kazakhstan, a U-2 pilot spotted something on the horizon and within a week the missile base at Tyuratam was reproduced as a scale model at the CIA. Photo intelligence had allowed the CIA analysts to determine the size and power of the Soviet rockets by noting the configuration of the launch pads and the size of the burn marks on the concrete.

The officer responsible for preparing National Intelligence Estimates at CIA, Ray Cline, briefed CIA Director Allen Dulles in early May 1957, but the intelligence was decided to be "too technical" for the National Security Council. Dulles changed his mind suddenly at the NSC meeting of May 10th, putting Cline on the spot to explain the "disturbing new evidence about Soviet weapons." President Eisenhower asked him if an ICBM launched from the Kamchatka peninsula could reach the United States. Cline responded by warning that Hawaii would be within range but not California.

The recent mushrooming of the VANGUARD budget continued at an alarming rate; so much so that both the Department of Defense and the National Science Foundation were becoming reluctant to continue the funding to completion. The CIA contributed $2.5 million from the Director's discretionary fund to keep the project going in the short term. At the NSC meeting, Dulles warned of the propaganda consequences of a US abandonment of the IGY satellite program. "If the Soviets succeeded in orbiting a scientific satellite and the United States did not even try to, the USSR would have achieved a propaganda weapon which they could use to boast the superiority of Soviet scientists. In the premises, the Soviets would also emphasize the propaganda theme that our abandonment of this peaceful scientific program meant we were devoting the resources of our scientists to warlike preparations instead of peaceful programs." Dulles recommended that the VANGUARD should be continued despite it being massively over budget. The reason was neither the future intelligence applications nor the establishment of "freedom of space" but plain old national prestige.

For those people with meager budgets, VANGUARD could have been built for a fraction of the cost, in fact, the cost of the hobbyist book *Discover the Stars* and the materials to construct the model. That was just one of the many books and magazines promoting the VANGUARD project at a time when its future was actually under threat. The Soviet Union was also promoting its satellite program in the press, promising a launch "within the next few months." The Soviet journal *Radio* published two articles in June, which gave detailed explanations of how *Sputnik* could be tracked by amateurs on frequencies of about twenty and forty megahertz.

As Soviet delegates to IGY committees and conferences became more vocal about the proximity of their launch, Korolev was having serious problems with the R-7 rocket. The first three launches of the ICBM in May-July 1957 were all failures, denying him the opportunity to orbit a satellite before the official start of the IGY. He too, like Milton Rosen, faced cancellation of his satellite project.

Von Braun continued to work on the nose cone testing for the US Air Force, using the ABMA Redstones, misleadingly designated the Jupiter-Cs. The first had already been a record-breaking success but the second had gone off course in May. Now in August, the third Jupiter-C performed perfectly. The nose cone survived re-entry and the flotation devices used in the water landing survived the attentions of sharks. Von Braun publicized this nose cone as "the first manmade object ever recovered from outer space." It was an important development in the protection of nuclear warheads, but it also brought an end to the involvement of the Huntsville team. With the successful re-entry trial, the Jupiter-Cs were no longer required and ABMA's contract with the US Air Force was ended. General Medaris ordered von Braun to keep two of the remaining Redstones fully assembled and ready for use if the call suddenly came to launch a backup.

On August 21st, 1957, on the fourth attempt, Korolev launched an R-7 over a record-breaking distance of four thousand miles into

Kamchatka. The Soviet Union broke with tradition and announced this successful launch to the West six days afterwards through the official news agency TASS.

> A few days ago a super-long-range intercontinental missile was launched. The tests of the missile were successful; they fully confirmed the correctness of the calculations and the selected design. The flight of the missile took place at a very great hitherto unattained altitude. Covering an enormous distance in a short time, the missile hit the assigned region. The results obtained show that there is the possibility of launching missiles into any region of the terrestrial globe. The solution of the problem of creating intercontinental ballistic missiles will make it possible to reach remote regions without resorting to strategic aviation, which at the present time is vulnerable to modern means of antiaircraft defense.

The TASS announcement was meant to frighten the West, and this was certainly the effect on its press. But Eisenhower had been aware of the R-7 before its launch. Tyuratam had been photographed again by the renewed U-2 flights, this time out of the secret CIA base in Lahore, Pakistan, during the three week long Operation SOFT TOUCH. It has been claimed that one of the U-2 sorties had observed the R-7 on its launch pad. But, despite efforts to make the U-2 less visible to radar than before, the spy plane was still tracked by the Soviets.

September 17th, 1957, was the 100th anniversary of the birth of the spaceflight visionary Konstantin Tsiolkovsky, and Korolev gave a long speech in his honor at the Pillard Hall of the Palace of Unions in Moscow. He announced to the audience that the first test launches of satellites would soon take place in both the USSR and the US. He would rather have celebrated the centenary with a launch from Tyuratam, but, despite a second successful launch of an R-7 ICBM in early September, the *Sputnik* program had slipped slightly in its schedule. A provisional target date of October 6th was set by the State Commission, who also decided to announce the launch of the satellite publicly only after the completion of its first orbit.

Korolev's concerns about an imminent US launch were increased by reports that an IGY conference was to be held in Washington in early October, with the US delegation to present a paper on October 6th, entitled "Satellite Over the Planet." He believed that this was deliberately timed to coincide with a US launch attempt. Despite KGB assurances that this was not the case, Korolev pressurized the State Commission into moving forward the date two days to October 4th.

The conference at the National Academy of Sciences in Washington opened on September 30th, with delegates from the Soviet Union joining those from the United States and five other nations to discuss protocols for sharing the scientific data and findings. In the opening session, Sergei Poloskov announced in a paper entitled "*Sputnik*" that now, "on the eve of the first artificial Earth satellite" the transmitters in the Soviet satellite would broadcast on frequencies of twenty and forty megahertz. These were in contravention of the stipulated IGY standard of 108 megacycles agreed the year before. Since the American radio tracking stations were about to go on line the very next day, the Soviets were asked when they planned to launch. Adjustments to the frequencies would have to be made and this would take time and money, so the American delegation pushed for a more exact date. To the laughter of the audience, Poloskov tried to avoid the question several times, but the surprise disclosure of the "eve of launch" was reported on the front page of the next day's *New York Times*.

A few days later, on October 4th, the newly designated Secretary of Defense, Neil McElroy, visited Huntsville and the Redstone Arsenal. He was on an orientation tour before being sworn in to replace Charles Wilson. Von Braun welcomed the appointment of McElroy since Wilson had been universally hated by the ABMA team. McElroy arrived at Huntsville around noon with a large entourage from Washington, which included the Secretary of the Army and the Chief of Staff of the Army. Major General Medaris and von Braun aimed to make a renewed pitch to the incoming Defense Secretary for an ABMA satellite attempt using the Jupiter-C.

Meanwhile, in Tyuratam, at exactly 22:28:34 Moscow Time, the mighty R-7 engines ignited to generate nearly four hundred tons of thrust. Korolev watched as his booster rocket lifted Tikhonravov's satellite off the launch pad and into history. When reports came in from Kamchatka that the *Sputnik* was passing overhead, the crowd at Tyuratam started to cheer, but were immediately cut off by Korolev. The Chief Designer warned: "Hold off on the celebrations. The station people could be mistaken. Let's judge the signals for ourselves when the satellite comes back after its first orbit around the Earth." Ninety minutes after launch, the distinctive beep-beep of the *Sputnik* could be heard over the radio and the crowd began to celebrate. Korolev learned that the R-7 had suffered a slight engine cut-off problem, which would result in the *Sputnik* failing to achieve its desired orbit. As a consequence, the satellite was losing altitude and would not stay in space as long as Korolev had hoped. State Commission chairman Ryabikov waited until *Sputnik* flew over Tyuratam for a second time before telephoning Khrushchev.

By the third pass over the Western Hemisphere, the *Sputnik* signal had been picked up by an RCA receiving station at Riverhead, New York, and relayed to the NBC radio studio in Manhattan where it was recorded for rebroadcast. The IGY scientists, after the first day's conference sessions, moved to the Soviet Embassy for an evening cocktail party. There Walter Sullivan, a science reporter for the *New York Times,* was suddenly called to the telephone. His Washington editor informed him that a bulletin had just been received from Moscow announcing the launch of a Soviet satellite. Sullivan, who had already filed a story that evening for the coming weekend's issue, saw his suspicions confirmed. A new report would have to be written. He walked back into the cocktail party and approached one of the key figures in the US IGY Committee, the chairman of the technical panel for placing an American satellite into space. "It's up," he whispered into the man's ear.

Although the man had suspected for some time that the Soviet launch was imminent, the news still came as a shock. He had been intimately involved with the American space program since its very early days and he felt responsible for its failure to win the race. He thought of VANGUARD and he thought of ORBITER. If only von Braun had been allowed to provide the booster rocket, everything would have been different. The chairman had been responsible for bringing von Braun into the United States in the first place. He was also responsible for the under-utilization of von Braun's talents in the early days of Project HERMES. The chairman of the technical panel was, of course, Dr Richard Porter.

The bad news was passed on to Dr Lloyd Berkner, President of the US IGY Committee, who clapped his hands and called for silence to announce: "I've just been informed by the *New York Times* that a Russian satellite is in orbit at an elevation of nine hundred kilometers. I wish to congratulate our Soviet colleagues on their achievement." The group of international scientists let out a cheer and Berkner proposed a toast. Vodka flowed and many guests moved out onto the Embassy roof, attempting to see *Sputnik* with the naked eye. The Soviet Embassy itself opened its normally closed doors to a multitude of reporters. One of the disheartened American members, the chairman of the US National Committee of the IGY, put on a brave face and was quoted as describing the *Sputnik* launch as "fantastic". He had cause to feel responsible for the failure of the US space program. His name was Joseph Kaplan and he had been one of the members of the secret Stewart Committee who had voted against von Braun's ORBITER.

The other man responsible at the Soviet Embassy cocktail party was his fellow secret Stewart Committee member and anti-von Braun voter, Dr Richard Porter.

Sputnik's Triumph

Von Braun was also at a cocktail party when the *Sputnik* bombshell exploded. He and Major General Medaris were entertaining McElroy and his entourage after an afternoon's tour of the Redstone Arsenal at Huntsville. Gordon Harris, the public affairs officer at the base, broke the bad news to Medaris: "General, it has just been announced over the radio that the Russians have put up a successful satellite. It's broadcasting signals on a common frequency and at least one of our local 'hams' has been listening to it."

According to Medaris' own account in his biography, *Countdown to Decision*, von Braun "started to talk as if he had suddenly been vaccinated with a Victrola [gramophone] needle. In his driving urgency to unburden his feelings, the words tumbled over one another."

"We knew they were going to do it!" exclaimed von Braun. "VANGUARD will never make it. We have the hardware on the shelf. For God's sake, turn us loose and let us do something. We can put up a satellite in sixty days, Mr McElroy! Just give us a green light and sixty days!"

During dinner, the Secretary of Defense designate sat between General Medaris and von Braun and received the hard sell on ABMA and the Jupiter-C. As an example of the criticism that Huntsville had for the present Secretary of Defense, Medaris pointed out to McElroy that more than a year earlier, von Braun could have launched a Jupiter-C nose cone into orbit if it hadn't been for Secretary Charles Wilson. Such was the hatred for Wilson among the Huntsville team that someone strung up a hanging effigy of the Secretary of Defense in the town.

"Listen now," said the NBC radio announcer, "for the sound that forever separates the old from the new." Next came the deep beep-beep of the *Sputnik*, a chirp lasting three-tenths of a second followed by a similar timed pause, then repeated over and over again. Two stories are worth telling of political stars of the future, one of whom was, as yet, still over the horizon. Senate majority leader Lyndon B. Johnson heard of *Sputnik*'s launch while hosting a barbecue at his Texas ranch. After dinner, he took his guests on a long walk and during the stunned silence, he philosophized: "As we stood on the lonely country road that runs between our house and the Pedernales River, I felt uneasy and apprehensive. In the open West, you learn to live with the sky. It is a part of your life. But now, somehow, in some new way, the sky seemed almost alien." Fully aware of the great political opportunity that presented itself, Johnson wasted no time in phoning his Senate colleagues of both parties to get their support for investigative hearings on the matter.

Meanwhile, the young US senator from Massachusetts, John F. Kennedy, was drinking with his brother Bobby in a men-only bar at Boston's Loch Ober Café. The future President of the United States was deep in argument with an MIT professor, Charles "Doc" Draper, about the meaning of the *Sputnik* launch. Kennedy, the man who would eventually initiate the race to the moon, had very different views back in October 1957. He argued with Draper that all rockets were a waste of money and their use in space even more so.

With two future presidents showing opposite reactions to the *Sputnik* launch, it was ironic that the then current president was apparently showing no reaction to the crisis at all. Eisenhower was relaxing at his golf course retreat in Gettysburg.

McElroy spent the night at Huntsville, which allowed von Braun and

Medaris to renew their attack the next morning. They promised that they would launch America's first satellite within ninety days, a more realistic time period imposed by Medaris over von Braun's over-enthusiastic sixty days. "When you get back to Washington and all hell breaks loose," von Braun said, "tell them we've got the hardware down here to put up a satellite anytime."

Once McElroy and his entourage left Huntsville, Medaris and von Braun immediately swung into action as if the official permission had already been granted. Medaris gave the orders to get the Jupiter-C rocket No.29 out of mothballs and ready to launch. The Major General later admitted that this unauthorized action was a court-martial offense, but it was assumed that the order would eventually be given and time was running out fast. Von Braun's hatred for Secretary of Defense Wilson and everyone in Washington who had denied him and his rocket team the opportunity to run in the space race, was so strong that he couldn't resist the opportunity to hit back publicly. On the Saturday morning of October 5th, immediately following McElroy's departure, a "scientist asking that his name not be used told the Associated Press that he was 'angry and distressed' because the Redstone Arsenal could have had a satellite into space before Sputnik if only it had been given the IGY assignment in 1955."

On October 5th, Washington's fears that the Soviet Union would capitalize on the propaganda value of launching the first satellite ahead of the Americans seemed to be unfounded. The initial Pravda article was exceptionally low-key and not even the headline of the day. Positioned modestly in a right hand column part way down the front page, its title was "TASS REPORT." The article gave the basic information on the Sputnik and paid homage to the great Konstantin Tsiolkovsky, who established the feasibility of artificial satellites at the end of the nineteenth century. The Soviet delegation at the IGY Conference in Washington became media celebrities and the propaganda benefits were once again held in check. Anatoli Blagonravov was asked on the NBC-TV show Youth Wants To Know if Sputnik was a victory over the West.

"We did not consider it necessary to compete in this field and we would be happy, no less than we are happy now, if we see the American satellite in space. We believe that our satellite, as well as the American satellite, could do it and serve science." However, Blagonravov had a different comment to make away from the cameras, in the US National Academy of Sciences. When given the floor to speak about *Sputnik*, Blagonravov criticized the United States for talking so much about its VANGUARD satellite before putting it in orbit. America should have adopted the Soviet approach of doing something first and then talking about it later.

Americans present found little comfort in the fact that IGY was all about disseminating information in advance of the launches and that the Soviet approach was not in the spirit of science. Joseph Kaplan's word of praise was quoted in the next day's *Pravda*, which devoted almost the entire front page to the story, under the head-line "WORLD'S FIRST ARTIFICIAL SATELLITE OF EARTH CREATED IN SOVIET NATION." There were no comments from Korolev, Glushko, or Tikhonravov. None of them had been publicly acknowledged as having any connection with the program; officially, they did not exist, so they could not comment.

TASS handled the press releases announcing both the short- and long-term objectives of the Soviet space program. More satellites would be launched within the IGY period, they said, with each new satellite heavier and more instrumented than the last. The IGY satellites would "pave the way to interplanetary travel and apparently our contemporaries will witness how the freed and conscientious labor of the people of the new socialist society makes the most daring dreams of mankind a reality." With no real official response, the US press fed on the press releases of the Soviets during the weekend.

By Monday October 7th, Lyndon B. Johnson had already arranged with Senator Richard Russell of Georgia, Chairman of the Senate Armed Services Committee, to have the Defense Preparedness Subcommittee, over which Johnson chaired, undertake an inquiry into why the Soviets had beaten the Americans into space. Russell was fully aware that his protégé Johnson had a competitor in the shape of Missouri Senator Stuart Symington, who would attack the President on party political lines. To Russell, this would cause trouble for the Democrats in that Symington's criticisms would not be in the national interest. Therefore, Russell backed Johnson's subcommittee proposal and avoided Symington's full-blown Senate Armed Services Committee inquiry proposal.

President Eisenhower returned from his weekend of golf to start dealing with the crisis, asking Donald Quarles, now the Deputy Secretary of Defense, to explain exactly why the United States was in second place. Quarles was probably the most experienced man to answer such a question, having been the Deputy Secretary of Defense for Research and Development during the crucial period of the Stewart Committee selection of the VANGUARD over the ORBITER. Quarles explained to the President that a decision had been made to completely separate the civilian scientific satellite program from the military one in order that foreign scientists working under the auspices of the IGY could be denied access to defense secrets. Looking on the bright side of the defeat, Quarles suggested that the Soviets had probably done the United States a favor in that they had established, beyond doubt, the concept of freedom of international space. Eisenhower, of course, had heard all this before, during the National Security Council meeting that had issued the directive NSC 5520. The *Sputnik* had crossed over many sovereign states and not a single one had objected, thus eliminating the legal problems feared by the Administration.

Eisenhower was particularly annoyed at the reports he had read of the comments made by two army officers concerning VANGUARD and ORBITER, and asked Quarles about the charges made. Eisenhower was referring to an Associated Press report on the criticisms made by Major General Holgar Toftoy and Brigadier General John Barclay, both of the Redstone Arsenal at Huntsville. Toftoy was, it must be remembered, the Army Ordnance officer responsible for acquiring the V-2s and their designers from Nazi Germany back in 1945. He and Barclay were attending the International Astronautical

Congress in Barcelona, Spain, and both chose to vent their anger against the Eisenhower Administration's denial of an army satellite launch program. These generals from Huntsville had used an international forum to launch an attack on VANGUARD and the President.

Quarles replied to Eisenhower that there was no doubt that the German-designed Army Redstone rocket could have orbited a US satellite a year before *Sputnik*. Later that morning, Eisenhower met with the outgoing Secretary of Defense, Charles Wilson. According to his autobiography, the President instructed Wilson to have the Army prepare its Redstone at once as a backup for the navy VANGUARD. Wilson, for whatever reasons, chose to forget to pass this on.

The Defense Department made two recommendations to the President. First, there should be no sudden change in the VANGUARD satellite program as a result of the *Sputnik* launch. The Soviets had beaten them into space and nothing would change that fact. Second, the President should issue a public statement that the United States had chosen to divorce its scientific IGY satellite from its military missile program, unlike the Soviet Union. *Sputnik*, as the Defense Department reasoned, was of little use scientifically. In fact it was superior to von Braun's original 5lb inert sphere, but the President could and should mention the more advanced instrumentation on the American VANGUARD. In response to anticipated fears that the *Sputnik* would raise doubts about the state of the US military missile program, the President should stress that he had chosen to develop the nation's missiles in an entirely separate program

By October 9th, *Pravda* finally published comments from Korolev, Glushko, and Tikhonravov—all anonymously, of course. Two of them acquired titles as pseudonyms: Korolev became known to the Soviet people as the Chief Designer of Rocket-Space Systems; and Glushko was the Chief Designer of Engines. Tikhonravov must have been rather peeved at not being allocated a title. The Soviet Union's desire to keep their identities secret was explained by Khrushchev: "In order to ensure the country's security and the lives of these scientists, engineers, technicians, and other specialists, we cannot yet make known their names or publish their photographs." Korolev would not be named publicly until after his death in 1966. In 1957, his name was removed from official histories of Soviet rocketry and encyclopedias now listed him as heading a laboratory in a "machine building" institute in the USSR. All three leading characters and their real names were well known to Western Intelligence thanks to the information provided by the German debriefings of Operation DRAGON RETURN.

President Eisenhower carried out two ceremonies in the White House on the day of his major press conference. In the first, he presented the Medal of Freedom to outgoing Secretary of Defense Wilson, and in the other he swore in Neil McElroy as Wilson's successor. After the ceremony reception was over, Eisenhower called all of the top Pentagon officials into the Oval Office for a dressing down. The President was still furious at the antics of generals Toftoy and Barclay in Barcelona. "When military people begin to talk about this matter and to assert that other missiles could have been used to launch a US satellite sooner, they tend to make the matter look like a race, which is exactly the wrong impression. I want to enlist the efforts of the whole group on behalf of, 'No Comment' on this development."

The President also had to deal with the White House press corps on the same day. He asserted that the United States had not been in a race with the Soviets to be first into space. *Sputnik* was only "one small ball in the air," but through it the Soviets had certainly "gained a great psychological advantage through the world." Perhaps, in hindsight, the United States could have tried harder to be first into space, Eisenhower felt. He admitted that the psychological advantage gained by the Soviets had been discussed beforehand but did not seem to be a reason to get "hysterical about it." The launching of *Sputnik* certainly demonstrated that the Soviets possessed rockets capable of launching nuclear warheads thousands of miles, but he added that there was no evidence that they had solved the problems of accuracy and re-entry. Concerning Khrushchev's comment that missiles would soon make bombers obsolete, Eisenhower argued that there would be a long transitional period of "evolutionary, not revolutionary" development. "So far as the satellite itself is concerned, that does not raise my apprehensions, not one iota."

This comment was meant to calm down the American hysteria that was developing over *Sputnik* and what it meant to US national security. But Eisenhower, through his apparent lack of concern, only raised public anxieties. Having waited almost a week for their president to give his response to the crisis, the American people started to question his ability to lead. He was a great military man who was expected to deal with the military matters raised by *Sputnik*. Instead, he seemed strangely uncertain and confused to the American public. A glaring example of this was his reply to a question from journalist Robert Clark who wanted to know how the Russians had gotten ahead in launching a satellite. According to Eisenhower, the reason was: "from 1945, when the Russians captured all of the German scientists in Peenemünde...they have centered their attention on the ballistic missile."

This statement was an extraordinary one to make and has never really been explained. The *Huntsville Times* Washington correspondent wrote, "By his words at his weekly press conference, the President indicated that he was unaware of the world-famous team of former German scientists now at the Army's Redstone Arsenal at Huntsville, Alabama." Von Braun was a television celebrity through his participation in Walt Disney's documentaries on space travel. It was common knowledge that German scientists had been brought to the United States at the end of the war. Operation PAPER-CLIP had caused uproar among the Jewish and scientific communities in America. It was also known that some of the German scientists had gone to work for the Soviets and they were still assumed to be there, since the mass repatriation of Germans from the Soviet Union had not been widely covered in the press. The association of German rocket scientists with the space programs of both the United States and the Soviet Union was well enough known for Bob Hope to react live on television to the launch of *Sputnik* with that witticism, "Well, let's not get too upset. That simply means that the Russians' Germans are better than our German scientists."

But according to Eisenhower there were no "our German scientists." Neither Eisenhower nor his press secretary James Hagerty would comment on the matter. Was it possible that the President was making a derogatory remark about the Huntsville team? Certainly the commanding officer of Redstone Arsenal Major General Toftoy and his deputy, Brigadier General Barclay, had greatly annoyed Eisenhower. A scientist from Huntsville "asking that his name not be used" had criticized the Department of Defense for not supporting the Redstone Jupiter-C launch program. Would that have irritated Eisenhower enough to make him pick specifically on the Germans? It seems unlikely. Was the comment meant as a slur on the Soviets? Was Eisenhower devaluing the Soviet achievement by attributing it to the Germans instead? That would make more sense. Eisenhower had discussed with his scientific advisors the need to not "belittle the Russian accomplishment" in his public response, but maybe this was just too hard to resist. When one focuses exactly on what Eisenhower said, a new picture emerges.

"From 1945, when the Russians captured all of the German scientists in Peenemünde..." is actually correct. One must ignore the Huntsville Germans because von Braun and his team surrendered themselves in the Bavarian Alps, having been transferred from Peenemünde by the SS. When the Soviet Red Army captured Peenemünde in May 1945, there were no leading scientists, engineers, or technicians left there. Any who were nearby were of no importance to either von Braun or SS General Kammler who had selected the five hundred to be transferred to the Final Redoubt. Therefore if the Soviets captured any of the ones remaining in the Peenemünde area, they probably captured all of them. This explanation would make Eisenhower unbelievably pedantic and suggests a degree of preparedness in his response, as opposed to an off-the-cuff remark that was actually in error. Eisenhower would repeat the remark in a later television appearance, but scaled down the number of German scientists from "all" to "most." This shows that it was still his purpose to falsely attribute the *Sputnik* success to non-Russians, as a slur on the Soviet technological capability. It also implies that Eisenhower was never going to allow Germans to put an American satellite into space first either.

Whatever the reasons for Eisenhower's comment, it was not well appreciated by von Braun and his team down in Huntsville. A "top missile man whose name is nationally known" was quoted as saying that the Huntsville team's offers to help the VANGUARD program had been refused but they could launch their own satellite in as little as three weeks. Although von Braun was a civilian, he was still under contract to the US Army, and as such, was subject to the official gag order from Eisenhower. So if he was the well-known missile man, and there is little reason to doubt it, then he was only making his case worse by turning up the heat.

VANGUARD project director John P. Hagen had been in regular contact with President Eisenhower during the crisis, trying to convince him to stick with the VANGUARD. There had been three test firings already and the fourth was due to take place in December 1957. It would be live and Hagen would be making an attempt to launch VANGUARD into orbit. Since the scheduled satellite launch was supposed to be the sixth in the series, he cautioned against any public announcement in case it failed. Launching rockets was a very risky business and important events were likely to backfire if too much attention was drawn to them. It was just one of those natural laws.

The Soviets had announced the launch of Sputnik only after it was already in orbit. With such a closed society as the Soviet Union it was entirely possible that the Sputnik launch was not the first attempt. Unless Hagen was privy to secret intelligence gleaned from the CIA or NSA monitoring stations, he would have to believe that either the Soviets had been lucky to get a first time launch or they had covered up a previous failure or failures. Hagen's cautiousness was brushed aside when Eisenhower's press secretary James Hagerty boldly announced on Friday 11th October that the next scheduled VANGUARD launch in December would carry the IGY satellite.

Over the weekend, the parallel secret American satellite burst onto the scene. The United States needed to let the world and, more importantly, its own people know that it hadn't been sitting on its hands for years. *Aviation Week* disclosed that the US Air Force was working on a multimillion-dollar program to develop a reconnaissance satellite. The WS-117L was accurately described and the technical problems of transmitting pictures back to earth were discussed in the article. The story was picked up by the *New York Times* on the Monday under the front page headline "US WORKING ON SATELLITE THAT COULD FILM THE EARTH—PROGRAM IS REPORTED UNDER WAY SINCE EARLY 1956—FIRST UNMANNED 'MOON' WOULD USE TV OR REGULAR CAMERAS."

Even this unauthorized disclosure of the secret USAF program could not allay fears of the advanced technology of the Soviets. On the same day as the *Aviation Week* issue came out on the news stands, Washington columnist Stewart Alsop was suggesting that the Soviets had already put a reconnaissance satellite into orbit. He wrote: "There is a mounting body of evidence, taken most seriously in the Washington intelligence community, that the Soviet satellite is not blind. *Sputnik* has eyes to see." Another fear was that the *Sputnik* beeps were encoded data being sent back to Earth and the CIA, NSA, and other intelligence services around the world were desperately trying to decipher them. One of the IGY delegates in Washington tried to calm things down by announcing that there was no coded signal and that the satellite could not see, but these assurances fell on deaf ears.

On October 21st, *Life* magazine published an article "WHY DID US LOSE THE RACE? CRITICS SPEAK UP," written by Clifford Furnas. Having served as Assistant Secretary of Defense for Research and Development from December 1955 to February 1957, he was well qualified to comment on the IGY satellite program and, more importantly, the voting decision of the Stewart Committee of which he had been a member. There had been nine members, of whom three had been what he described as "fence-sitters." These details of the secret committee would prove to be very misleading, since the numbers given were wrong.

He blamed the failure of the US satellite program on the "tragically naïve and short-sighted outlook" of the Department of Defense which regarded research "as a sort of extracurricular scientific pastime to be indulged in only if money is left over from the 'really important' things." The blame was squarely placed on the shoulders of the ex-Secretary of Defense, Charles Wilson. Furnas revealed that he had warned Wilson in 1956 that the United States would suffer a terrific propaganda defeat if it did not spend more money on the satellite program. Wilson's reply, according to Furnas, was "So what?"

Wilson's public statements were no less dismissive. "Nobody is going to drop anything down on you from a satellite while you are asleep, so don't worry about it," had been his soothing words while still Secretary of Defense. On his retirement a few days after *Sputnik*, Wilson was having trouble sleeping. His former aide, Colonel James George, found him just sitting in his house in Detroit, staring into space. Mrs Wilson and Colonel George eventually snapped him out of it. "I don't know why they're mad at me," he said, referring to the reports in the papers that his effigy had been hanged at the Army Missile Base in Huntsville. "I'm the one who put them in the Jupiter business. They must have me mixed up with the Ku Klux Klan...It's only natural when things go wrong for Americans to look for a fall guy or a goat. If some of them want me to be the goat, that's all right. If I were a smart man I would never have taken the job in the first place."

Thanks to the media publicity about the thwarted attempts by the Huntsville rocket team to be first into space, von Braun became a national star in his own right. Even his shady past was being looked at in a new light. In a *New York Times* magazine interview that appeared in October, he told the story about SS Reichsführer Heinrich Himmler and speeding up the production of V-2s. "I remember he was interested in horticulture, and I said that it was all right to put a plant into the ground and carefully nurture it, but that too much manure, for example, could kill it. Himmler smiled weakly, but I thought little of it until I was arrested three weeks later by the Gestapo. It took a direct order from Hitler to release me."

A direct order from Eisenhower was now required to release the Jupiter-C from its mothballs. The new Secretary of Defense McElroy met with the President on October 30th, and proposed that the ORBITER be used as a backup to VANGUARD, at an additional cost of $4 million. Eisenhower agreed, adding angrily that he had wanted such a step taken eighteen months previously but Wilson and the Department of Defense had stressed that the civilian and military programs should be kept separate. Wilson's negligence over passing on the President's recent order to prepare ORBITER must have frustrated Eisenhower even more. But delays under McElroy would be just as bad. It would be another week before the official go-ahead was publicly given. In the meantime, the Soviet Union blasted a second *Sputnik* into orbit on November 3rd, 1957, and gave the world its first cosmonaut—albeit not a human one.

"It appears that the name Vanguard reflected the confidence of the Americans that their satellite would be the first in the world," declared Khrushchev in a speech at the 40th anniversary of the Revolution on November 6th. "But...it was the Soviet satellite which proved to be ahead, to be in the vanguard...in orbiting our earth, the Soviet *Sputniks* proclaim the heights of the development of science and technology and of the entire economy of the Soviet Union, whose people are building a new life under the banner of Marxism-Leninism." As he watched a parade of tanks, rockets, and missiles in Red Square, he boasted further, "Our satellites are circling the earth waiting for American satellites to join them and form a commonwealth of *Sputniks*."

After the launch of the first *Sputnik*, Korolev had gone to the Kremlin and Khrushchev told him: "We never thought that you would launch a *Sputnik* before the Americans. But you did. Now please launch something new in space for the next anniversary of our revolution." With the 40th anniversary of the Revolution only one month away, Korolev told everyone in his design staff to be "guided by his own conscience" for there would be no time for quality checks. Khrushchev's demand for "something new" would be satisfied by the inclusion of a mongrel dog named Laika aboard *Sputnik* 2.

When the second satellite was launched on November 3rd, the sheer size of it stunned American scientists and missile analysts. Weighing slightly more than 1100lbs, it was six times heavier than the first, and dwarfed the six-inch diameter of the VANGUARD. Scientists estimated that the rocket used to launch *Sputnik 2* must have had a thrust of at least 500,000lbs—more than enough to power an ICBM a distance of five thousand miles. Democrat senator Stuart Symington claimed that the awesome feat of the latest launch proved that the United States was at least two years behind the Soviet Union in ICBM technology. The infamous "missile gap" became a growing political misconception that would ultimately lose Richard Nixon the 1960 election and power John F. Kennedy into the White House.

British scientists suggested that the Soviets had produced a new chemical fuel or perhaps even a nuclear engine. If so, then it would not be long before the Soviet Union launched a rocket to the moon. The *New York Times* carried a front-page story that foreign scientists believed that a Russian rocket was already on its way to the moon, carrying a hydrogen bomb. It would detonate in a flash that would be brighter than the light of a full moon. Crazy speculations aside, the American public appeared to be more concerned with the fate of the dog, Laika. Although dismissed by Secretary of State John Foster Dulles as "a real circus performance," the presence of the dog on board the satellite angered animal rights protest groups, and raised fears among the American IGY committee that the Soviets were gearing up for manned spaceflights. Laika would eventually die in space and be portrayed by the Soviets as a martyr for a noble cause.

Eisenhower also suffered from the effects of re-entry. From a post re-election high in the opinion polls of seventy-nine percent, he had seen twenty points burned off, thanks to the *Sputnik* crises. In an attempt to recover his popularity, he appeared on a television and radio broadcast from the White House on the evening of November 7th. "I am going to lay the facts before you—the rough and the smooth," he warned at the start. Admitting that the Soviets had scored an important victory with the *Sputniks*, he went further and stated that if the technology were used in missiles carrying nuclear warheads it would "damage us seriously." Overall, he said, the military strength of the Western world was still greater than that of the communist countries, and the Soviet missiles were no deterrent against the destructive retaliatory power of the Strategic Air Force.

In an answer to doubts about his Administration's spending on the new missile technology, Eisenhower announced that more than $1 billion a year was going into missile research and development. The US Air Force had already tested ICBMs up to a distance of three and a half thousand miles, he added, and a team of scientists and engineers had solved the problem of re-entry. "One difficult obstacle on the way to producing a useful long-range weapon is that of bringing a missile back from outer space without its burning up like a meteor because of the friction with the earth's atmosphere." Pointing to the large white nose cone positioned on the floor beside his desk in the Oval Office, he continued, " Our scientists and engineers have solved that problem. This object here in my office is an experimental missile—a nose cone. It has been hundreds of miles to outer space and back. Here it is, completely intact."

Eisenhower's speech ended with a reminder that what the world needed "even more than a giant leap into outer space...was...a giant step towards peace." He would not be happy until a time when the scientist could "give his full attention not to human destruction, but to human happiness and fulfillment." The televised broadcast received a lukewarm reaction in the press, with one columnist describing it as "another tranquility pill." Von Braun and the Redstone Arsenal were fuming at the unattributed use of their Jupiter-C nose cone. But their period of frustration was almost at an end. Almost. The day after Eisenhower's "chins-up" speech, and thirty-five days after the first *Sputnik* was launched, the Department of Defense issued the following press release:

> The Secretary of Defense today directed the Department of the Army to proceed with launching an earth satellite using a

modified Jupiter-C. This program will supplement the VAN-
GUARD project to place an earth satellite into orbit around the
earth in connection with IGY. All test firings of VANGUARD have
met with success, and there is every reason to believe VAN-
GUARD will meet its schedule to launch later this year a fully
instrumented scientific satellite. The decision to proceed with
the additional program was made to provide a second means of
putting into orbit, as part of the IGY program, a satellite which
will carry audio transmitters compatible with Minitrack ground
stations and scientific instruments selected by the National
Academy of Sciences.

The Assistant to the Secretary of Defense for Guided
Missiles, Mr W.M. Holaday, will be responsible for coordinating
the army project as part of the United States satellite program.

This press release differed from the actual official directive sent
to Huntsville. They were aware that the statement "all test firings
of VANGUARD have met with success" was incorrect but accepted
it as part of the general propaganda. What they were not willing to
accept was the replacement of "to proceed with launching" by "to
prepare to launch." Medaris saw this as denying his team the right
to launch, and telephoned Bill Holaday at once. He reluctantly
agreed that that was the correct interpretation of the orders.
Medaris exploded at this point, feeling that the Huntsville team
deserved the right to launch anyway, irrespective of the outcome of
the scheduled VANGUARD launch in December. If they could not
get permission to go ahead, they wanted to quit and go on to other
work. Medaris, with the assent of both von Braun and the satellite
producer, William Pickering of the Army's Jet Propulsion Laboratory,
put it on a wire to General Gavin, the chief of Research and
Development of the Army, with a copy to the Secretary of the Army.

Medaris got an answer the next day from General Lyman
Lemnitzer, Army chief of staff, who acted as though he was unaware
of Medaris' outburst, but was troubled by the outburst coming from
von Braun. The President had been angered by the interview given

by von Braun to Associated Press representatives on the West Coast in which he had spoken out rather bluntly about the need for a speedier response. This was going against the President's public statement that there was no race with the Soviets, and the head of Associated Press had to be persuaded to censor the inflammatory comments. Lemnitzer warned Medaris to get out of the headlines and exercise control over von Braun and his damaging mouth.

No sooner had Medaris given his word to do this than, on November 14th, he and the troublesome von Braun travelled to Washington to brief defense officials and the press on their plans. At a packed news conference, the General announced that there was a ninety percent chance that the army would put a 21lb bullet-shaped satellite into orbit on the first attempt. Columnist Drew Pearson had written a recent article on the "six satellites," all ready to launch, that were gathering dust in Huntsville. These were not satellites, but the Jupiter-C rockets, so Medaris was telling the truth when he denied their existence. The satellites were in fact held in a locker at the Jet Propulsion Laboratory in Pasadena. Although ordered to cancel the ORBITER project, the JPL had continued to develop their satellites in parallel with the Redstone Arsenal's Jupiter-Cs. Those involved in the deception called it "bootlegging" and the ORBITER satellite was renamed DEAL, after the poker term. When asked at the press conference if work had already started on the army satellite, Medaris admitted that it had but gave the impression that it had only just started.

It was hard to stay out of the headlines as the VANGUARD launch date of December 3rd approached, as well as the opening of Lyndon B. Johnson's hearings before the Senate Preparedness Subcommittee on November 25th. The senator from Texas compared the *Sputnik* crisis to Pearl Harbor, even ranking it a greater challenge since "we do not have as much time as we had after Pearl Harbor." Despite this sense of urgency, the hearings would last nearly two months, with seventy-eight witnesses giving over 2,300 pages of testimony. The first witness called was Edward Teller, the "father of the H-bomb" who spoke in ominous terms of the implications of *Sputnik*.

It was a clear sign that the Soviets were beginning to take the lead in science and technology, he said, adding that the Americans had waited too long before starting their missile program, thus giving the Soviets a head start which proved to be too much in the race into space. Any rocket powerful enough to launch a satellite the size of *Sputnik 2* was capable of being an ICBM.

Vannevar Bush, the man responsible for mobilizing American scientists in World War II, stressed the need for a greater emphasis on basic scientific research rather than a crash weapons program, and criticized the inter-services rivalry that he saw as the root of the missile program failure to beat the Soviets. World War II hero James Doolittle concentrated on the basic need for a higher priority in teaching science at school level. VANGUARD was defended by its project director, John Hagen, who explained to Johnson that it could have beaten *Sputnik* into space if only it had received higher priority. Such a request was made in 1955, he said, but he had received no response from the Department of Defense. Half way through Hagen's testimony, Johnson interrupted him to publicly announce to the subcommittee and assembled press that President Eisenhower had just suffered a heart attack.

The excuse given for his non-attendance at the dinner that evening at the White House in honor of the visiting King of Morocco, was that he had suffered a chill after greeting the monarch at the airport. According to his doctor, Howard Snider, it was more likely that Eisenhower had suffered a slight heart attack that only affected his speech patterns. Eisenhower was to make a full recovery, but in the meantime Vice-President Nixon agreed to stand in at the state dinner, much to the annoyance of the President.

After Johnson had made the announcement, instead of postponing the hearings, he allowed Hagen to continue. At one point, a light bulb from the overhead chandelier fell down onto the committee table and Johnson asked if the bulb was part of his VANGUARD project. Hagen replied, "No sir, I think that it is one of those strange flying objects." Hagen did his best to defend VANGUARD, but since he had only come aboard the project after its selection, he was not qualified to comment on that phase. That role should have been assigned to a man who sat in on the hearings but was never called to testify.

Over the first couple of days of the hearings, the Advisory Group on Special Capabilities, which had selected VANGUARD, was discussed with no reference to the name by which it had unofficially been known. Over the first couple of days of the hearing, the Stewart Committee's decision to select the VANGUARD was discussed but with reference to the official title of the Advisory Group on Special Capabilities. Several attempts were made by witnesses to remember the name of its Chairman, Homer Stewart. Because of its secret nature, no-one helped out. What was even more intriguing was the fact that not only was Stewart present at the hearings, keeping quiet about his chairmanship of the secret committee, but he was also on the other side of the tables, acting as the senior scientific advisor to Johnson's Armed Services Subcommittee. Sitting behind the senators, his job was to suggest questions and interpret testimony for them. Evidently it did not involve answering any questions himself.

In fact, none of the members of the Stewart Committee were summoned to testify at the hearings. Clifford Furnas would have been an ideal witness, with his experience of both the Stewart Committee selection procedure and his later role as Assistant Secretary of Defense for Research and Development. There may be some credence to the theory that he was ignored because of the bipartisan stance he took in the article which appeared in *Life* after the launch of *Sputnik* 1. Although he had criticized the Eisenhower administration for its inefficient handling of the satellite program, he had also laid initial blame on the preceding Truman administration for letting the Soviet Union get such a head start. Although Johnson claimed his hearings were bipartisan, they were far from it. There would be no criticisms of the Democrat's missile policy and Truman's missile manager Kaufman Keller was never called, even when von Braun made an attack.

The only representatives of the Eisenhower administration called to give testimony on missile research and development were

Donald Quarles and William Holaday, the latter having no link with the IGY program at all. Quarles and his boss, Secretary of Defense McElroy, had to support Eisenhower's defense spending budget ceilings against accusations that these were responsible for the limited resources supplied to the missile development. Both men claimed that the $38 billion ceiling was only interpreted as a guideline and that money had always been redirected from lower priority programs to ensure sufficient funding for the missiles. On the second and third days of the hearings, CIA Director Allen Dulles and nuclear arms expert Herbert Scoville testified in a top-secret, closed session before the committee, scaring the senators with pessimistic short-term and long-term outlooks on the perceived missile gap.

General James Gavin revealed the extent of the army's fears when he told the senators that, after being turned down five times, Medaris and von Braun were so concerned that the Soviets were going to go into space first that they discussed the possibility of shooting down the first Russian craft. There would be no way of knowing whether the satellite was civilian or military, and something would have to be done to prevent such an intrusion of American airspace. Bearing in mind the NSC's obsession with the legal concept of "freedom of space," such action would have been disastrous.

The chairman of the IGY committee and another former Stewart Committee member not called to testify, Joseph Kaplan, held a press conference in Chicago on December 2nd, only three days before the scheduled launch of VANGUARD. He cautioned reporters about the "risk of failure" but assured them that before the end of the International Geophysical Year, the United States would have "a full-fledged earth satellite in orbit." If the launch in three days' time failed, that allowed VANGUARD another twelve months to get it right.

With the world's press assembled at Cape Canaveral on the Atlantic coast of Florida in anticipation of the launch of America's first satellite, the flight's postponement due to technical problems came as a monumental embarrassment. The American public had no experience of the failure rates of rockets so could only view the delay at Cape Canaveral in the worst way. With two *Sputniks* already up in orbit, it was only a question of time before the Soviets rubbed American noses further in the dirt by launching a third. Secretary of State John Foster Dulles exploded in rage at the next day's National Security Council meeting, calling the delay "a disaster for the United States" that had made their country "the laughing stock of the whole free world."

If that was how he felt about a mere technical delay, how would he feel about something worse?

The VANGUARD Conspiracy

With the whole world watching, the VANGUARD satellite lifted off at precisely 11:44:55am on Friday December 6th, 1957. Two seconds later, the rocket was buckling under its own weight and bursting into flames. With an element of black comedy, the 3.2lb, six-inch diameter VANGUARD somehow managed to escape from the collapsing wreckage and roll into the scrub brush where it was later found beeping.

Both Hagen and Eisenhower had been spared the agony of seeing the disaster, with the project director listening to the countdown by phone in Washington and the President doing likewise from his retreat at Gettysburg. But not for long, as the film footage of the launch-pad explosion was shown on television over and over again. The world's press had a field day thinking up humorous names to call the pathetic VANGUARD. The *New York Times* dubbed it "Sputternik," the *New York Herald Tribune* called it "Goofnik," *Time* magazine referred to the whole program as "Project Rearguard," the *News Chronicle* labelled it "Stayputnik," the *Daily Express* "Kaputnik", but it was the British *Daily Herald* that caught the eye of the propaganda hungry *Pravda*. There were two photographs, one of the VANGUARD being readied on the launch pad and the other of the explosion, underneath the large headline "OH, WHAT A FLOPNIK!" Superimposed on the *Daily Herald* front page were the *Pravda* comments, "*Reklama*" and "*Deistvitelnost*"—"Publicity" and "Reality."

One person who did not find that at all funny was Korolev. He knew how difficult it was to launch a rocket and Soviet ridicule of

VANGUARD would only bring down a curse upon his next project, *Sputnik* 3, the original OBJECT D. For that reason, he complained to Khrushchev, and the *Pravda* campaign was stopped immediately. However, the Soviet delegation at the United Nations could not resist asking their American counterparts if the United States would be interested in receiving aid under their program of technical assistance to backward nations.

The reasons for the spectacular explosion were soon investigated, thanks to the extensive filming of the event and the two seconds-worth of onboard telemetry before the explosion. Did the fault lie in the first stage engine, the X-405, or the liquid propellant tank? The two components were manufactured by two different companies, and technical investigators from each were desperate to blame the other for the failure.

The propellant tank manufacturer was the Glenn L. Martin Company and their experts traced the explosion to an "improper engine start" which led directly to a low fuel tank pressure and a resultant low fuel injector pressure. This allowed some of the burning contents of the thrust chamber to enter the fuel system through the injector head, they claimed. With the fuel injector destroyed, there was a complete loss of thrust immediately after lift-off.

The other company disagreed with their findings. They denied there was "an improper start" and blamed the explosion on a loose fuel line connection, leaking fuel on top of a helium vent valve which blew down onto their engine. There were accusations that members of the Glenn L. Martin Company had used the fuel lines as ladders while working on the rocket, and the other company claimed that this had loosened the connections.

The VANGUARD project technical director, Milton Rosen, accepted the Martin findings, but persuaded the other company to alter the minimum allowable fuel tank pressure head of its engine by thirty percent in an effort to speed up the next launch date. The other company accepted the change in specifications without accepting the blame. The X-405 had not been on the original Viking rockets. It was added to the VANGUARD booster after it had been selected by the Stewart Committee. What would interest conspiracy theorists was the fact that the X-405 was built by the General Electric Company and had been developed from the old Project HERMES engine designed by Dr Richard Porter, a key member of the Stewart Committee.

The original manufacturer of the Viking engines had been Reaction Motors Inc., who were working on a new thrust of 75,000lbs when Milton Rosen and the Naval Research Laboratory switched to the 27,000lbs thrust of the GE X-405. Investigative reporters Drew Pearson and Jack Anderson asked Porter about the fact that his company General Electric benefited from the switch in motors. His response was: "I don't know anything about the contract." In a subsequent conversation with the reporters he apologized for having been evasive previously. "I knew the Vanguard would be using GE engines when I voted for it, but this did not influence my decision. Our recommendation was made strictly on the basis of technical factors."

Pearson and Anderson built up a case against Porter. They researched his past involvement with the V-2 rockets in Germany and at White Sands, New Mexico, and his employment with GE and the army during the acquisition and debriefing of the von Braun team. As chief engineer, Porter had designed the Hermes engines that General Electric installed in the A-1, A-2, and A-3 missile series. There were reports that Porter's improvement on the V-2 motor was not an improvement at all, and others claimed that, because of the trend towards smaller warheads, those missiles soon became obsolete. For whatever reasons, the US Army cancelled the series. With the von Braun team moving to the Redstone Arsenal in Huntsville, Alabama, General Electric were left holding the bag in New Mexico.

As time went on, Porter rose in the astronautical ranks until he was the president of the American Rocket Society. The chairman of the society's Space Flight Committee was Milton Rosen who just so happened to be in charge of the Viking rocket program. Pearson and Anderson claimed that Porter sold Rosen on the idea of developing an improved Viking with a GE motor and using it as a satellite launcher while they were in charge of the American Rocket Society. A technical report was written by Rosen, and Porter forwarded it on to the National Science Foundation, which was preparing for the International Geophysical Year.

The IGY Committee, headed by Joseph Kaplan, was so impressed by the feasibility study that it invited Porter to become head of the satellite panel. When the then Assistant Secretary of Defense for Research and Development, Donald Quarles, selected members of his Advisory Group on Special Capabilities, he decided on Dr Homer Stewart as its chairman with Porter and Kaplan from the IGY Committee. The other members of the nine-man team were representatives of the three armed services. Quarles ensured that two men were nominated from each service with a non-voting secretary in the shape of Admiral Paul Smith. The final decision would rest with Quarles, on the advice of the Stewart Committee. Pearson and Anderson had harassed Stewart for details of the committee which bore his name.

"I can't talk. My instructions don't permit me to comment at all."

"You can't even comment as to whether you are chairman of the Stewart Committee?"

"No. I can't comment on whether I am chairman of the committee. All I can say is that I've served on several committees."

When the Johnson hearings resumed in mid-December, the Texan senator soon issued a press release which stated that the country's problems with rockets and missiles were not caused by men such as VANGUARD's John Hagen, but by the men at the top who could not make "hard, firm decisions." A Gallup opinion poll of December 15th laid most of the blame for the nation's "missile gap" with the Soviets on the President and the Republicans. In second place was inter-service rivalry.

This type of rivalry also existed within the Stewart Committee which had been deliberately set up to reflect the three services' interests in the IGY satellite selection. With a project from the air force, ORBITER from the army, and VANGUARD from the navy, and two members nominated by each service, it would be no surprise if the members voted along the lines of service affiliation, in which case the deciding votes would be cast by the two Quarles' nominees. Two of the voters were also leading members of the IGY satellite committee and it was obvious that they would have a disproportionate amount of influence. These two men, of course, were Dr Porter and Dr Kaplan. But who were the other members of this secret committee that rejected von Braun's ORBITER and backed the ultimate loser VANGUARD?

Johnson's Senate Preparedness Subcommittee failed to find out who these members were. With Stewart advising the senators at the hearings, was it any wonder? It was up to the House Armed Services Committee and its "Investigation of National Defense Missiles" in January and February 1958 to reveal their identities. In addition to Homer Stewart, Joseph Kaplan, Richard Porter, and Clifford Furnas, the voting members were Charles C. Lauritsen, eminent physicist and professor at Cal Tech, George H. Clement from the RAND Corporation, J.Barkley Rosser, rocket ballistician and professor of mathematics at Cornell University, and Robert McMath, professor of astronomy and head of the McMath Hubert Observatory at the University of Michigan.

Determining which members were nominated by which service is a guessing game, but one which can be played with a degree of certainty. Stewart and Furnas both had strong army connections and both have gone on record as voting for the army's ORBITER project. Clement worked for RAND, which was a US Air Force think tank, and Rosser was affiliated to it also. Lauritsen had been involved in navy rocket development since 1940, but who was the other navy nominee? It would have to be either Porter or McMath, since Kaplan, as chairman of the US National Committee for the IGY, was obviously appointed by Quarles. McMath had been suffering from an illness during the selection process and had been unable to attend the crucial voting session. Afterwards, he let it be known that he would have voted for ORBITER. Why his vote could not have been carried out by proxy is never explained. Anyway, his preference for ORBITER rules out any likelihood of a navy affiliation, which leaves Porter as the navy nominee. Having seen the General Electric HERMES project cancelled by the army, Porter was unlikely to have any affiliations with the army despite his activities in wartime Germany. Pearson and Anderson pointed out the links between Porter and Milton Rosen over the incorporation of the GE X-405 engine into the VANGUARD project and this would explain Porter's vote for VANGUARD. Kaplan had already shown a preference for the navy project before the Stewart Committee was even formed.

If McMath had been allowed to vote during his absence, the ORBITER would have had three members supporting it. Three other members, Porter, Kaplan, and Lauritsen, voted for VANGUARD, which left two fence-sitters. These were the presumed air force nominees, Clement and Rosser. Neither man was enthusiastic about ORBITER and VANGUARD, which they saw as marginal in payload and unlikely to meet the IGY launch deadline. The air force's entry in the IGY stakes was the half-hearted attempt known as the "World Series," which eventually became an unrealistic proposal. With their RAND background, neither Clement nor Rosser would have veered from the official RAND advice to the US Air Force that a small scientific satellite was a waste of time. With their own service project abandoned, they would have been under great pressure to conform to USAF guidelines and not vote for the "enemy", i.e. the army. The rivalry between the air force and the army was far greater in the field of missiles—where some of the same components were used in rival rocket engines—than that between the air force and the navy. Nevertheless, there were two "fence-sitters" who, it was claimed, had little experience in rocket engineering and therefore fell in line with the majority vote. With McMath's absence, the vote was two for ORBITER (committee chairman Stewart and Furnas) and three against (Porter, Kaplan, and Lauritsen). Clement and Rosser then voted with the majority and VANGUARD was selected.

If only McMath's vote had been allowed, there would have been no majority for Clement and Rosser to follow and the outcome may have been different. Pearson and Anderson maintained that the "fence-sitters" had been pressurized by Porter to vote against ORBITER because he had a commercial interest in the VANGUARD project. The journalists went on to publish their accusations in a 1958 bestseller *USA—Second Class Power?* They claimed that Porter's advice to the Stewart Committee had been tainted by corrupt motives to secure a major contract for his employers General Electric. The heavy investment made in Project HERMES needed to be utilized after the army cancelled the program.

The allegations were investigated in August 1959, at the Special Investigations Subcommittee at the House Armed Services Committee. Porter testified that he had offered to step down from the Stewart Committee when the VANGUARD proposal came before it, but Stewart himself had persuaded him to stay. (This account was never confirmed by Stewart.) Porter also produced a letter from VANGUARD director John Hagen attesting to the reliability of the General Electric X-405 engine. This "unsolicited" testimonial was rumored for many years afterwards to be the product of GE's public relations department, and not Hagen. Porter's appearance at the 1958 investigations hearings was in stark contrast to his absence at the Johnson hearings. He would come to represent what Eisenhower hated the most and coined in his farewell address, the "military-industrial complex."

But could there have been another reason for the Stewart Committee's decision to reject ORBITER? Could there be something not related to inter-service rivalries or the relative merits of the satellite payloads? Was there an anti-German conspiracy within the Committee? One of the originators of the International Geophysical Year and a member of the ORBITER group, S. Fred Singer, wrote to Wernher von Braun on August 24th, 1955, expressing dismay at the "rather antagonistic feelings" held by "members of the IGY group" against von Braun and the Redstone Arsenal.

The chairman of the US National IGY Committee, Joseph Kaplan, did not want the American entry to be tainted by the Nazi connection. The Redstone was a development from the German V-2 and that would have been too great a liability in the international arena.

As far as Kaplan was concerned, von Braun and his Germans from Peenemünde were an arrogant group who saw themselves as the best rocket scientists in the world. Their arrogance had constantly hackled Dr Richard Porter, ever since he first met them at Garmisch-Partenkirchen in 1945. Their overwhelming air of superiority had continued in the deserts of New Mexico during Project HERMES, when Porter's missiles failed to impress. Kaplan and Porter were the dominant force within the US IGY Committee and they ended up being the same within the Stewart Committee. There was a definite meeting of minds with Milton Rosen, and VANGUARD represented the real "American" satellite project. The new American citizens from Huntsville were still the Nazis from Peenemünde.

Von Braun was fully aware of the anti-German prejudice against him and his team. But he was not alone. Helmut Gröttrup had suffered it too in Russia. Korolev and his Soviet designers had deliberately isolated the German workforce on the island of Gorodomlya. They had no idea whether their design plans were ever used by the Soviets in their missile program. The repatriations in the early 1950s had left them with a professional sense of uselessness. There was no evidence of a German contribution to the success of *Sputnik*; its great scientific achievement was down to the advances made by the Soviet engineers. Korolev wanted to get into space with a Soviet rocket, not a modified German one. In this way, he was just the same as the US IGY Committee wanting to get into space with an American rocket. The arguments about civilian over military rockets clouded the issue. VANGUARD's Viking rocket belonged to the US Navy, and as such was a military rocket. When the authorities talked about not utilizing a military rocket, what they really meant was a *German* military rocket.

And yet, it was von Braun and the Germans who came to the rescue of their adopted country.

In a further attempt to demilitarize the Redstone rocket, the Jupiter-C had been renamed the Juno. In Roman mythology Juno was the sister and wife of Jupiter. The rocket would be bearing the child of Jupiter, the satellite renamed DEAL.

As the Juno waited patiently on Pad A of Launch Complex 26 at Cape Canaveral, the VANGUARD rocket continued to have problems. On January 22nd, President Eisenhower was asked by an aide if he wanted to issue a lengthy statement the next day if VANGUARD was successfully launched. He replied that he would prefer a brief White House press release simply stating that the first in a series of scientific experiments had been conducted successfully. Other members of his party were more eager to cash in on the success, with several of them prerecording speeches to be released when the satellite entered orbit.

On January 23rd, the first attempt to launch failed because heavy rains shorted some of the ground instrumentation cables during the countdown. The next three days saw three more countdowns, two of them coming to within seconds of firing before being cancelled. On one occasion, with the launch held at 22 seconds, a newsman shouted to the assembled reporters to look upwards. Rather ominously, Sputnik 2 passed overhead. Finally, on January 26th, the VANGUARD crew had to call off the attempts until a second stage engine could be replaced. The Huntsville team was given five days to launch their Juno before priority would return to VANGUARD.

With the Juno ready to launch on January 29th, there was an unfortunate delay due to the jet stream shifting southwards. This meteorological phenomenon consisted of a high-velocity "river of air" at altitudes of between twenty-five thousand and forty-five thousand feet with sharp pressure changes at the upper and lower edges. It was these edges, which could shear the rocket, that worried Medaris and caused him to postpone. The weather reports the next morning were not much more encouraging. The window of opportunity would close after January 31st, and the launch was rescheduled for 10:30 that evening, one and a half hours inside the deadline.

As the jet stream headed northwards and out of range of the launch, Eisenhower was being updated on the weather reports at

his cottage next to the Augusta National Golf Course. Press secretary Hagerty agreed to let him know personally when the launch
was successful. If, for whatever reason, it had to be "scrubbed out"
he was to pass on a two-word message: "Nothing doing." Eisenhower
would be spending the night playing bridge and didn't want to be
interrupted unnecessarily.

The Germans at Cape Canaveral were led by Kurt Debus. Von
Braun, much to his anger and frustration, was at the Pentagon with
JPL's Pickering and Professor Van Allen who had installed an impor-
tant experiment on board the satellite.

At 9:45pm, someone spotted that hydrogen peroxide appeared
to be leaking out of the tail. The countdown was held and engineers
went to check it out. After draining the peroxide, there was no further leak and the countdown resumed fifteen minutes later. At
exactly 10:48pm, the firing ring was pulled and the Juno blasted off
sixteen seconds later with an earthshaking roar and an orange ball
of flame. From zero to 18,000mph would take approximately seven
minutes, but the time taken to confirm its position in orbit would
take an excruciating 100 minutes.

The confirming signal came in from the Army Tracking Station
in Earthquake Valley, California: "Goldstone has the bird!" The first
American satellite, renamed EXPLORER 1, was about to make its
first flight over the United States.

"That's wonderful," Eisenhower told Hagerty on hearing the
news. "I surely feel a lot better now...Let's not make too great a hul-
labaloo over this."

Von Braun, Pickering, and Van Allen rushed over to the press
conference at the National Academy of Sciences on Constitution
Avenue and celebrated the launch of EXPLORER by jointly hoist-
ing a model of the satellite above their heads. The triumphant
image would appear in newspapers across the country and across
the world. America had joined the space race, despite what
Eisenhower claimed.

Epilogue

Eisenhower and von Braun met for the first time at a formal white tie dinner at the White House a few days later to celebrate the launch of EXPLORER. Von Braun, so the story goes, rented a tuxedo for the event and to his dismay found that it came with a standard black tie. Using his new found status, he contacted Press Secretary Hagerty and asked for help. Not to worry, he was told, a white tie would be waiting for him on his arrival at the White House. When President Eisenhower made his belated entrance at the function, he was visibly annoyed at having to wear a black tie since, as he complained to Hagerty, "I can't seem to find my white tie anywhere."

The two men met again in Washington on January 17th, 1959, when President Eisenhower presented him with the Distinguished Federal Service Medal, the highest award given to non-military personnel. There must have been a certain amount of déjà vu for von Braun. After a lengthy period of political indecisiveness, chronic under-funding and denial of opportunity, Hitler had awarded von Braun that War Service Cross when the V-2 was successfully launched. Now, fifteen years later, President Eisenhower had done more or less the same thing. There are other even more striking parallels. Inter-service rivalry between the German Army and the Luftwaffe's V-1 project had led to the fatally late deployment of the V-2 after the Allies had already entered Europe. Inter-service rivalry between the American Army and the US Navy's VANGUARD project had led to the fatally late deployment of the Jupiter-C booster after the Soviets had already entered space.

At the beginning of this book, reference was made to Eisenhower's comment that if the V-2s had been deployed six months earlier, operation OVERLORD would have been cancelled. It raises the question about alternative histories, the intriguing "What if?" speculations on the different outcomes of events at the major turning points of history. Hitler himself admitted that the V-2s, if deployed in large numbers, would have ruled out the need for future wars. If the Reich had won the war, it is very probable that von Braun would have reached the moon a lot sooner than he did.

What if von Braun and his team had been allowed to launch that Jupiter-C in 1956, with a satellite on board the fourth stage? For a start, there would have been no panic over the Soviet missile capability, which *Sputnik* produced. Many people, in Washington and NATO, equated satellite launches with ICBM prowess. If the Soviets were capable of putting *Sputniks* of increasing weight into orbit, they were perceived as more than capable of launching nuclear warheads at the United States. With the apparent inability of the United States to respond immediately to the *Sputnik* launches, a very real fear was generated that there existed a technological gap between the superpowers. The satellite launching feats indicated that there was a "missile gap", with the Soviets well in the lead. This gap had to be explained and blame attributed. Eisenhower's military defense budget ceilings were immediately criticized, even though his spending on missiles far surpassed that of his predecessor, Harry S. Truman. The Democrats jumped at the opportunity to paint the former general as someone who had allowed US military prestige to fade. The missile gap, backed by US Air Force and CIA intelligence estimates, became a crucial factor in the 1958 Congressional elections which resulted in an increased share for the Democrats.

Senator John F. Kennedy was particularly vocal about the missile gap, having changed his mind about the significance of rockets since that *Sputnik* night argument with an MIT professor in Boston. Gaining the Democratic nomination ahead of the likes of Stuart Symington and Lyndon B. Johnson, Kennedy went on to criticize the Republican nominee Richard Nixon over his party's neglect on defense. In the presidential election of 1960, Kennedy won by less than sixty thousand votes, leading some political commentators to question whether the result would have been the same if the missile gap had not been a factor. If von Braun had beaten Korolev into space, the world could very well have been a different one. No missile gap panic. No President Kennedy. No Cuban missile crisis.

On January 31st, 1961, eleven days after the real inauguration of President Kennedy and exactly three years after von Braun's team had launched EXPLORER I, a Mercury-Redstone rocket blasted off from Cape Canaveral, carrying a chimpanzee named Ham. Technically speaking, he was the first American in space and was the first sign that the United States had not only caught up with the Soviets but was now taking the lead. Up until then, Korolev had only managed to launch dogs into orbit.

A lot had happened to the German rocket scientists in the intervening three years. They were still at Huntsville but had been transferred out of the Army Ballistic Missile Agency and were now at the core of the civilian National Astronautics and Space Administration (NASA) at the Marshall Space Flight Center. Von Braun ran the Marshall Center with the help of his deputy, Dr Eberhard Rees, who had been at his side since the days of Peenemünde. In fact, the heads of department were dominated by names from the old days and the place had a definite Germanic feel about it. The bulk of the PAPERCLIP boys took up contracts with NASA but some of the original team had moved into private industry, such as his brother Magnus (Chrysler Corporation) and Dr Ernst Steinhoff (RAND Corporation). Others had even returned to Germany (Walter Riedel and Hermann Oberth). Some old American Allies had left Huntsville with Generals Toftoy and Medaris and Colonel Hamill retiring from the Army Ordnance. Times were changing. The heroes of the space race would no longer be the rocket designers but the astronauts they helped to put into space.

With Alan Shepard set to replace Ham on the next Mercury-Redstone launch scheduled for March 24th, the prospect of America winning the race to get a man into space seemed sure. But Wernher

von Braun was unsure. Astronaut Ham had been severely traumatized by the short space flight. He had been exposed to far greater
gravitational forces than had been expected and von Braun needed
to test another Mercury-Redstone instead, without a passenger.
Ham was in no fit state to make another flight.

The Mercury-Redstone test flew perfectly and Shepard's attempt
was rescheduled for May 5th. The day after von Braun's test, Korolev
successfully launched and recovered the dog Zvezdochka on board
Sputnik 10. He was now finished experimenting with dogs; it was
time to put a human cosmonaut into space. On April 12th, 1961, the
very same day that the new NASA chief James Webb was due to
proudly show the House Committee on Science and Astronautics
old film footage of traumatized Ham's Mercury capsule recovery,
one of Korolev's R-7 rockets placed Yuri Gagarin in orbit. Alan
Shepard was understandably disappointed. "We had 'em by the
short hairs," he said, "and we gave it away."

For the second time Korolev had achieved a space first, and this
time von Braun had been involved in a fair race. His caution had
cost him the glory.

Kennedy was quick to congratulate the Soviets: "We, all of us,
as members of the race, have the greatest admiration for the
Russian who participated in this extraordinary feat." Privately, he
asked Vice-President Lyndon Johnson in his role as chairman of the
National Space Council: "Do we have a chance of beating the Soviets
by putting a laboratory in space, or by a trip around the moon, or by
a rocket to land on the moon, or by a rocket to go to the moon and
back with a man? Is there any other space program that promises
dramatic results in which we could win?"

Johnson replied that landing a man on the moon would have
"great propaganda value...In the eyes of the world, first in space
means first, period; second in space means second in everything."

Von Braun had told Johnson that the United States had "an
excellent chance of beating the Soviets to the first landing of a crew
on the moon." He pointed out that the Russians would need to build
a rocket engine ten times as powerful as the current one to achieve

the task, and the US was already three years into the development of the gigantic F-1 engine with its 680 tons of thrust.

Kennedy had learned from the *Sputnik* crisis and the VANGUARD failure the importance of controlling the publicity surrounding rocket launches. It was with great annoyance that he learned of an arrangement made between *Life* magazine and the previous Eisenhower Administration over publication rights of the first manned American launch. Even worse was the previously agreed television coverage of the event. Had the Republicans learned nothing from the VANGUARD disaster? If the launch failed and Shepard died in the tragedy, the public reaction would be horrific.

When broadcasters and the press were unmoved by presidential pleas to resist covering the launch live, it was down to Wernher von Braun and his team to safely put Shepard into a brief fifteen-minute suborbital flight and then bring him back down again. Once this was achieved, the resultant press coverage raised public interest in American space travel once again and this may have given Kennedy confidence to famously announce during a speech to the Congress on May 25th: "I believe that this nation should commit itself to achieving the goal, before this decade is out, of landing a man on the moon and returning him safely to earth."

Former President Eisenhower had very firm views on a manned mission to the moon. While still in office he had asked a panel of scientists to evaluate such a plan and they had believed it was worth doing. Eisenhower rejected the whole business as nothing more than a "stunt" and privately declared that he "couldn't care less whether a man ever reached the moon."

With an attitude like that, Eisenhower would be classified along with all the other leaders von Braun had struggled to convince. With Kennedy, at last, he had a political leader who shared his dreams.

Bibliography

Bower, Tom, *The Paperclip Conspiracy: The Battle for the Spoils and Secrets of Nazi Germany*, London: Paladin Grafton Books, 1988

Burrows, William E., *Deep Black: Space Espionage and National Security*, New York: Random House, 1986

Dickson, Paul, *Sputnik: The Shock of the Century*, New York: Walker and Company, 2001

Divine, Robert A., *The Sputnik Challenge*, Oxford: Oxford University Press, 1993

Dornberger, Walter, V 2, London: Hurst and Blackett, 1954

Green, Constance M., and Milton Lomask, *Vanguard: A History*, Washington: NASA, 1970

Gröttrup, Irmgard, *Rocket Wife*, London: Andre Deutsch Ltd., 1959

Hall, R. Cargill, 'Early U.S. Satellite Proposals' in Emme, Eugene, (ed.) *The History of Rocket Technology*, Detroit: Wayne State University Press, 1964, pp. 67-93

Harford, James, *Korolev*, New York: John Wiley & Sons, Inc., 1997

Hunt, Linda, *Secret Agenda: The United States Government, Nazi Scientists, and Project Paperclip, 1945 to 1990*, New York: St. Martin's Press, 1991

Lasby, Clarence G., *Project Paperclip: German Scientists and the Cold War*, New York: Atheneum, 1971

Launius, Roger D., John M. Logsdon and Robert W. Smith (eds.), *Reconsidering Sputnik: Forty Years Since the Soviet Satellite*, Amsterdam: Harwood Academic Publishing Group, 2000

Logsdon, John (ed.) *Exploring the Unknown: Selected Documents in the History of the US Civil Space Program*, Vol.1: Organizing for Exploration, DIANE Publishing Company, 2003

McDougall, Walter A., *The Heavens and the Earth: A Political History of the Space Race*, New York: Basic Books, Inc., 1985

McGovern, James, *Crossbow and Overcast*, London: Hutchinson & Co., 1965

Neufeld, Michael J., *The Rocket and the Reich: Peenemünde and the Coming of the Ballistic Missile Era*, New York: The Free Press, 1995

Ordway, Frederick I., and Mitchell R. Sharpe, *The Rocket Team*, London: Heinemann, 1979

Pearson, Drew and Jack Anderson, *USA—Second Class Power?* New York: Simon & Schuster, 1958

Siddiqi, Asif A., *Sputnik and the Soviet Space Challenge*, University Press of Florida, 2003

Taubman, Philip, *Secret Empire: Eisenhower, the CIA, and the Hidden Story of America's Space Espionage*, New York: Simon & Schuster, 2003

Von Braun, Wernher, 'The Redstone, Jupiter, and Juno' in Emme, Eugene, (ed.) The History of Rocket Technology, Detroit: Wayne State University Press, 1964, pp. 107-114

Index